TECHNICAL REPORT

Assessing the Tradecraft of Intelligence Analysis

Gregory F. Treverton, C. Bryan Gabbard

RAND NATIONAL SECURITY RESEARCH DIVISION

This research was conducted within the Intelligence Policy Center (IPC) of the RAND National Security Research Division (NSRD). NSRD conducts research and analysis for the Office of the Secretary of Defense, the Joint Staff, the Unified Commands, the defense agencies, the Department of the Navy, the Marine Corps, the U.S. Coast Guard, the U.S. Intelligence Community, allied foreign governments, and foundations.

Library of Congress Cataloging-in-Publication Data is available for this publication.

ISBN 0-8330-3958-X

The RAND Corporation is a nonprofit research organization providing objective analysis and effective solutions that address the challenges facing the public and private sectors around the world. RAND's publications do not necessarily reflect the opinions of its research clients and sponsors.

RAND® is a registered trademark.

Published 2008 by the RAND Corporation
1776 Main Street, P.O. Box 2138, Santa Monica, CA 90407-2138
1200 South Hayes Street, Arlington, VA 22202-5050
4570 Fifth Avenue, Suite 600, Pittsburgh, PA 15213
RAND URL: http://www.rand.org/
To order RAND documents or to obtain additional information, contact
Distribution Services: Telephone: (310) 451-7002;
Fax: (310) 451-6915; Email: order@rand.org

Preface

Most public discussions of intelligence address operations—the work of spymasters and covert operators. Current times, in the wake of September 11th and the intelligence failure in the run-up to the war in Iraq, are different.[1] Intelligence *analysis* has become the subject. The Weapons of Mass Destruction (WMD) Commission was direct, and damning, about intelligence analysis before the Iraq war: "This failure was in large part the result of analytical shortcomings; intelligence analysts were too wedded to their assumptions about Saddam's intentions."[2] To be sure, in the Iraq case, what the United States did or did not collect, and how reliable its sources were, were also at issue. And the focus of post mortems on pre-September 11th was, properly, mainly on relations between the Central Intelligence Agency (CIA) and the Federal Bureau of Investigation (FBI) and on the way the FBI did its work. But in both cases, analysis was also central. How do the various agencies perform the tradecraft of intelligence analysis, not just of spying or operations? How is that task different now, in the world of terrorism, especially Islamic Jihadist terrorism, than in the older world of the Cold War and the Soviet Union?

The difference is dramatic and that difference is the theme of this report. The United States Government asked RAND to interview analysts at the agencies of the U.S. Intelligence Community and ask about the current state of analysis. How do those analytic agencies think of their task? In particular, what initiatives are they taking to build capacity, and what are the implicit challenges on which those initiatives are based? Our charter was broad enough to allow us to include speculations about the future of analysis, and this report includes those speculations. This report is a work in progress because many issues—the state of tradecraft and of training and the use of technology and formal methods—cry out for further study. This report was long delayed in the clearance process. It has been updated and remains a useful baseline in assessing progress as the Intelligence Community confronts the enormous challenges it faces.

---

[1] See, in particular, the reports of the two national commissions that investigated the two failures. They are, respectively, National Commission on Terrorist Attacks Upon the United States (2004); and Final Report of the Commission on the Intelligence Capabilities of the United States Regarding Weapons of Mass Destruction (2005) (hereafter referred to as the WMD Commission Report). The Senate Select Committee on Intelligence was equally scathing about the October 2002 National Intelligence Estimate, concluding: "Most of the major key judgments . . . either overstated, or were not supported by, the underlying intelligence reporting. A series of failures, particularly in analytic tradecraft, led to the mischaracterization of the intelligence." "Report on the U.S. Intelligence Community's Prewar Intelligence Assessments on Iraq" (2004).

[2] WMD Commission Report, p. 3.

This research was conducted within the Intelligence Policy Center (IPC) of the RAND National Security Research Division (NSRD). NSRD conducts research and analysis for the Office of the Secretary of Defense, the Joint Staff, the Unified Combatant Commands, the defense agencies, the Department of the Navy, the Marine Corps, the U.S. Coast Guard, the U.S. Intelligence Community, allied foreign governments, and foundations. It will be of interest to a wide range of people concerned with influencing others—from covert operators, to public diplomacy specialists, to foreign policy planners, to those involved in attempts at informing in other realms of policy. Questions or comments can be directed to the principal author, Gregory Treverton at gregt@rand.org. For more information on RAND's Intelligence Policy Center, contact the Director, John Parachini. He can be reached by email at John_Parachini@rand.org; by phone at 703-413-1100, extension 5579; or by mail at the RAND Corporation, 1200 South Hayes Street, Arlington, Virginia 22202-5050. More information about RAND is available at www.rand.org.

Contents

Figures

Tables

Summary

"Analysis" in the U.S. Intelligence Community is definitely plural. It encompasses a range of styles, levels, and customers. It ranges from solving puzzles (such as whether Iraq had weapons of mass destruction—a question that could be answered definitively if only the United States had access to information that in principle was available) to framing mysteries (those questions that are future and contingent, which no information could resolve definitively). It would surprise many citizens to learn that the big "collectors," such as the National Security Agency or the National Geospatial Intelligence Agency, have more "analysts" than the Central Intelligence Agency.

The vast majority of what all those analysts do is current and tactical, more question-answering than producing deep understanding of critical issues. That tyranny of the immediate has become more entrenched, for a variety of reasons, not least that technology now permits the take from big national collection systems to be retrieved in time to help warriors on the battlefield. In our conversations, that tyranny was sometimes applauded—as providing policymakers, including the president, what they wanted—but more often bemoaned. However, it always was noted.

There is no shortage of analytic tools being created, inside and outside the Intelligence Community. But there are concerns about the connection between those tools and the needs of analysts. Too often analysts regarded the tool-builders as in a world of their own, building tools that analysts could not quickly master. As one analyst from a Service intelligence organization put it, analysts are imprisoned not by organizations or sources but, rather, by tools.

At the same time, the analytic community faces both opportunities and challenges in dealing with a large cohort of new, young analysts, who are computer-savvy and networked. They take for granted an ease of access to information that has been the opposite of the Intelligence Community's compartmentalization and "need-to-know." They can become the drivers of a sea-change in how the Community thinks about analysis and sharing. Or they will be lost to the Community.

It was plain in our informal survey of the Intelligence Community that every agency has a separate set of research priorities and product lines. These varying missions and products serve a range of customers, from the president and his immediate advisors and Cabinet members, to key military leaders charged with day-to-day actions to secure the lives of Americans worldwide, to state and local law enforcement officials engaged in the war on terrorism. This broad constituency drives needs for a wide range of activities in both research and development, training, and education that are a challenge to coordinate across the nation's entire intelligence enterprise. Yet none of the agencies knows much of what its colleagues do, still less works with them consistently in testing and validating analytic techniques or in training analysts.

Accordingly, we concluded that the establishment of a research agenda and a training and education curriculum with a Community-wide perspective is critical to future analytic tradecraft. It is all the more important now, given the creation, in December 2004, of a director of national intelligence; and the establishment of a National Intelligence University is a welcome first step. Also important is a common reference point for judging the tradeoffs among stakeholder pressures for the various analytic tasks—pressures that bear on the Community at large in different ways. Our research also identified shortfalls in analytic capabilities, methodologies, and skills, and it recommends actions to take to address these gaps as well as a strategy for meeting future challenges.

Table S.1 presents a summary set of recommended actions for the Deputy Director of National Intelligence for Analysis, DDNI(A), as well as for the Chancellor of the National

Table S.1
Summary Set of Recommended Actions

Establish DDNI(A) as a focal point to evaluate opportunity costs and assess "right balance" in analysis
— Collection-driven versus issue-driven
— Current reporting versus longer-term analysis
— In-house versus outsourced

Foster better integration of methods and tools for analysis
— Establish focal point to connect R&D and tool-building community (government and industry) to Intelligence Community analysts
— Develop minimum common tool set for community-wide use

Institute community-wide tradecraft training and education components
— Develop tradecraft curricula for community-wide use
— Institutionalize lesson-learning as process of performance improvement, not assessing blame

Get and keep the next generation of analysts
— Build partnerships with academia (e.g., Media Lab), industry (e.g., Futures Lab), and government (e.g., NRO/AS&T/FL) and link new hires
— Track promotion, retention, and erosion rates for new hires over decade
— Align training, incentives, processes, and metrics with performance

Innovate in analytic methods and data-sharing
— Promote a variety of experiments and field tests, mostly "inside the security fence," as demonstrations and validations
— Recognize that the nature of secrecy is changing

Evaluate the boundaries of all-source versus single-INT analysis
— End the distinction at mid and high levels of analysis; analysis is not distinguished by the number of sources
— Develop portfolio of "Day After" games, and other simulations, to nurture transitions

Rethink new kinds of intelligence, especially law enforcement
— Focus on usefulness, necessity of "domestic/foreign" divide
— Use gaming to explore gray areas

Intelligence University, to move toward a future analytic community with enhanced and more agile tradecraft that will be essential in addressing a fundamentally different and uncertain era of global challenges over the long term.

Since the December 2004 legislation and the major post mortems, a number of initiatives have moved in directions we recommended. Perhaps most important is the DDNI(A) position itself. Just as the Director of National Intelligence has the possibility to build authority commensurate with responsibility for the entire Intelligence Community, so the DDNI(A) has the opportunity to become a real hub for a Community-wide perspective on goals, training, and tradecraft in analysis. So, too, establishing a National Intelligence University, whose chancellor is also the Community's Chief Training Office, can provide a focal point for training, including training in analysis; creating the National Counterterrorism Center and other centers can shift intelligence, including analysis, toward an organization around problems or issues, not agencies or sources; building a Long Term Analysis Unit at the National Intelligence Council can lead away from the prevailing dominance of current intelligence; and forming a DNI-managed Open Source Center can be a seedbed for making more creative use of open-source materials, as well as, perhaps, developing a model for other initiatives in analytic tradecraft.

These are promising actions, but they are works in progress. Perhaps they can begin to change the attitude that lies behind specifics. For all the admonitions and exhortations, the national and Community leadership devalues intelligence analysis today. For all the language about the importance of intelligence analysis, data-sharing, fusion priorities, and the like, the price of doing better is seen as too high for the likely results. Now is the opportunity to change that attitude.

Acknowledgments

We happily acknowledge our debt to our colleagues in the project Glenn Buchan and Kevin O'Connell, who contributed to earlier drafts of this report. We also appreciate the detailed comments of our RAND reviewers, David Aaron and Richard Hundley, and those of our sponsor. Their efforts have made this report better, and they should be absolved from responsibility for any shortcomings that remain.

Acronyms

ARC	Analytic resources catalogue
ARDA	Advanced Research and Development Activity
AS&T/FL	Advanced Systems and Technology/Futures Laboratory, NRO
ASW	Anti-submarine warfare
ATP	Advanced Technology Programs
BENS	Business Executives for National Security
CIA	Central Intelligence Agency
CMS	Community Management Staff
COMINT	Communications intelligence
CTC	Counterterrorism Center
DARPA	Defense Advanced Research Projects Agency
D&D	Denial and deception
DCI	Director of Central Intelligence
DDNI	Deputy Director of National Intelligence
DDNI(A)	Deputy Director of National Intelligence for Analysis
DHS	Department of Homeland Security
DI	Directorate of Intelligence
DIA	Defense Intelligence Agency
DMPI	Desired mean point of impact
DNI	Director of National Intelligence
DoD	Department of Defense
DoE	Department of Energy
DS&T	Directorate of Science and Technology, CIA
DTRA	Defense Threat Reduction Agency
EELD	Evidence Extraction and Link Discovery
ELINT	Electronic intelligence
FBI	Federal Bureau of Investigation

FBIS	Foreign Broadcast Information Service
FFRDC	Federally Funded Research and Development Center
GD	General Dynamics
GERP	Global Expertise Resources Program
HSI	Hyperspectral imagery
HUMINT	Human intelligence
IARPA	Intelligence Advanced Research Projects Activity
IDA	Institute for Defense Analyses
IMINT	Imagery intelligence
INR	Intelligence and Research Bureau, State Department
INT	Intelligence collection service, as SIGINT for signals intelligence
IR	Infrared
IRS	Internal Revenue Service
ITIC	Intelligence Technology Innovation Center
JMIC	Joint Military Intelligence College
LANL	Los Alamos National Laboratory
LLNL	Lawrence Livermore National Laboratory
MASINT	Measurement and signature intelligence
NASIC	National Air and Space Intelligence Center
NCCOSC	Naval Command Control Ocean Surveillance Center
NCS	National Cryptologic School
NCTC	National Counterterrorism Center
NGA	National Geospatial Intelligence Agency
NGIC	National Ground Intelligence Center
NGO	Nongovernmental organization
NIAPB	National Intelligence Analysis and Production Board
NIC	National Intelligence Council
NIMD	Novel Intelligence from Massive Data
NIPF	National Intelligence Priorities Framework
NIST	National Institute of Standards and Technology
NIU	National Intelligence University
NRO	National Reconnaissance Office
NSA	National Security Agency
NSF	National Science Foundation
NSPD	National Security Policy Directive

ONI	Office of Naval Intelligence
ORD	Office of Research and Development
OSC	Open Source Center
PDB	President's Daily Brief
QIC	In-Q-Tel Interface Center
R&D	Research and development
RDT&E	Research, development, test and engineering
ReBA	Rebuilding Analysis
S&T	Science and technology
SIGINT	Signals intelligence
SMO	Support to military operations
STEP	Science and Technology Experts Program
TIA	Terrorism Information Awareness
UN	United Nations
USMC	U.S. Marine Corps
WGI	Within grade increase
WINPAC	Weapons Intelligence, Proliferation and Arms Control
WMD	Weapons of mass destruction

CHAPTER ONE

Introduction

U.S. intelligence analysts today are pushing the limits of their craft and finding a welcome reception from some of their most senior consumers. At the same time, they are also stretched and frustrated with the uncertainty of their mission and buffeted in the wake of the national investigations of intelligence failure before September 11th and before the Iraq war. The contrast partly reflects differences across analytic agencies, but it also reflects differences within them. And it also reflects tensions within individual analysts over what they do and how they add value. In analysis, as in other areas, the Intelligence Community remains something between a loose federation and an aspiration. Analysts from one agency are not hostile to those in other agencies; they are mostly ignorant of one another. The need for a focal point in analysis and analytic tradecraft is striking, and this need will only grow as the Community strives to be more "joint" in the wake of the December 2004 intelligence reform law and the creation of a director of national intelligence.[1]

The overarching generality about the U.S. intelligence analytic community today is that most of it is engaged in work that is tactical, operational, or current. By most accounts, the relative lack of longer-term analysis has long been bemoaned. In other words, most analytic resources and activities are dedicated to intelligence reporting instead of attempting to attain the "deep understanding" of our adversaries that constitutes analysis. Why is this the case? As we will discuss below, it is a function of the complex security environment, the nature of decisionmaker's needs, personnel practices, and the success of our technical collection activities.

Ironically, recent government actions have exacerbated this situation, perhaps unintentionally. The National Intelligence Priorities Framework (NIPF) ratified in National Security Decision Directive 26, for example, identifies 150 priority intelligence targets, countries, or issues, emphasizing a dozen or so. This framework represents an official sanction for not paying attention to issues associated with more balanced global coverage. If the emphasis on immediate reporting is sharper now, that is so because it is what many national intelligence consumers want (or, at least, it is what they get because they do not ask for longer-term analyses) and because, for the warfighters in particular, an abundance of military intelligence, essential to such day-to-day operations as force protection, is available from national means.

---

[1] Formally, the Intelligence Reform and Terrorism Prevention Act of 2004, available at www.fas.org/irp/congress/2004_rpt/h108-796.html (accessed January 4, 2005).

In this RAND project, we sought to review, assess, and make recommendations about the Intelligence Community's priorities for research and development and training and education that might lead to better analytic capabilities in the future. In essence, this report documents the current status of the analytic community. It identifies, in turn, issues and shortfalls in the analytic community's use of methods and tools and its ways of organizing and using its most important resource, human skills. It then portrays the issues that will shape the analytic community of the future, concluding with suggestions keyed to the issues identified above. For some issues, we make specific, actionable suggestions. Others, however, go to the heart of what intelligence will be, and we try to focus sharply on those issues in our discussion.

The report reviews available data. While the data are improving, the limitations of existing data—limitations recognized at senior levels—is the subject of a strong recommendation. We have reviewed data on the demographic profile of the Community in the analytic resources catalogue (ARC), which is updated quarterly and correlated with the National Intelligence Priorities Framework.

In addition, we have relied on detailed interviews conducted in 2003–2004, with three dozen leaders and analysts in the analytic community and also on information we gained during meetings with line analysts. In the interviews, we coded views about major themes, and we divided those interviewed roughly into the "national" analytic agencies (the Central Intelligence Agency's [CIA's] Directorate of Intelligence ([DI]), the Department of State's Intelligence and Research Bureau [INR], the Federal Bureau of Investigation's [FBI's]Office of Intelligence, the Department of Energy's Office of Intelligence, and the Department of Homeland Security's Directorate for Information Analysis and Infrastructure Protection), the "big collectors" (the National Security Agency [NSA], the National Geospatial Intelligence Agency [NGA], and the National Reconnaissance Office [NRO], and the military Intelligence Community (the Defense Intelligence Agency [DIA] and the service intelligence organizations).

Given the ongoing reorganization of intelligence, names have been a moving target, but we also conducted interviews at components then under the Director of Central Intelligence and now under the Director of National Intelligence (DNI)—the National Intelligence Council; the then–Community Management Staff, now office of the DNI; and the then–Terrorist Threat Integration Center, now National Counterterrorism Center. At the major analytic agencies, such as the CIA's DI, we interviewed analysts from a number of components. In addition, we also conducted interviews at technology units that support analysis, such as the CIA's Directorate of Science and Technology (DS&T), the CIA's In-Q-Tel, and the Pentagon's Defense Advanced Research Projects Agency (DARPA).

The next chapter portrays the Intelligence Community's analytic cadres as they are today, drawing on our interviews and observations. Chapter Three turns to the contribution that technology, concepts of operation, time management, and other research can make to tradecraft through an effective research and development (R&D) program, and Chapter Four discusses approaches to improve human capital throughout the Intelligence Community. We conclude with a vision of intelligence analysis in the future.

CHAPTER TWO

The Analytic Community Today

"Analysis" in the U.S. Intelligence Community is definitely plural. One size of analysis no longer fits all, if it ever did. Figure 2.1 illustrates the multiple components of the analysis cycle that are of concern today.[1]

The canonical cycle began with policymakers and military leaders, whose concerns would be turned, by collection target planning analysis, into taskings for the major collectors. The take from those collectors would then be processed at various levels, ultimately to be incorporated into all-source analysis, then disseminated back to policymakers and military leaders. As the figure demonstrates, the cycle notionally distinguishes between intelligence sources and the analytic processes that are used to transform the raw data from these sources into intelligence products.

The dotted lines illustrate that the cycle can be short-circuited, and often is. Information at various stages, including "raw" intelligence, gets passed to policymakers. In some cases, illustrated by the dotted line marked A, the processing of collection system data could be in large measure automated in advance, with no person in the loop at the time of collection. The first level of "processing" would then be done with one computer system or network talking to another. It would be what one practitioner calls "generating DMPIs" (desired mean points of impact) for weapons targeting. Target coordinates might be transmitted directly to an airplane's cockpit or to an unmanned aerial vehicle (in which case both sensing and striking might be automated).

This may be contrasted with the analytic cycle more typical in the 1990s, one that typically highlighted three forms of analysis—technical processing analysis, single discipline analysis, and all-source analysis. However, if the distinction between single discipline and all-source analysis ever made sense, changing technology and changing threats have blurred that distinction. What "all-source" analysts produce, such as items for the President's Daily Brief (PDB), often are "single-source" products in that the new information comes from a single intelligence source, although the analysts strive to put that new information in context.

Imagine a continuum from collection system outputs at one end to analytic challenges at the other. Somewhere along the continuum, there is a transition region where the data must be used to support multiple forms of analysis. The analytic continuum splits between solving

---

[1] This elaborated cycle is only a point of departure for the assessment. It is true and widely believed that if the canonical cycle ever existed, it does no more. The shortcuts and dotted lines in the elaborated cycle arguably have become more important over time. See, for instance, Treverton (2001), p.104ff.

Figure 2.1
An Illustrative Intelligence Analysis Cycle

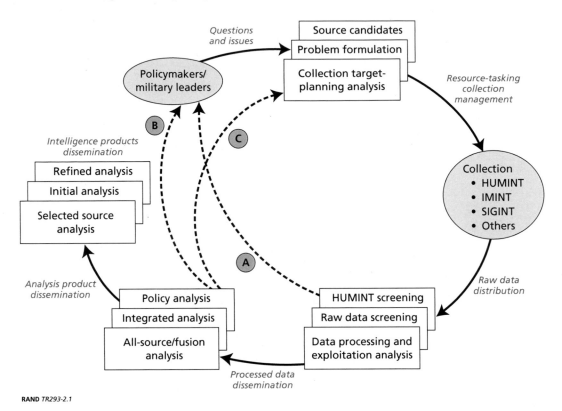

RAND *TR293-2.1*

puzzles and framing mysteries. Puzzles can be "solved" in principle if only we have access to information that does exist but may be unknown to us.[2] Many of intelligence's analytic successes in the Cold War were elaborate puzzles—how many missiles does the Soviet Union have, how accurate are they? Similarly, one of intelligence's signal failures in recent years was the elaborate puzzle of whether Saddam Hussein's Iraq had weapons of mass destruction (WMD) in 2003.

Scientific and technical analysis, the analysis and estimation of foreign system capabilities or R&D programs, were and are puzzle-oriented. Although those disciplines have waned, they remain mainline business for the NSA, the NGA, and the service intelligence agencies and centers, which seek to understand the weaponry of potential foes. Now, though, many, perhaps most, policy-level questions are mysteries, ones that are contingent on and depend in part on our actions, where intelligence can never provide definitive answers no matter how much information is collected. The answers are unknowable: When and how might al Qaeda strike the United States again?

At the other end of the puzzle-solving continuum would be complex puzzles akin to those of the Cold War but perhaps involving current threats and adversaries, such as terrorists, who

---

[2] This distinction is also made widely, in somewhat different forms, in discussions of analysis. See, for instance, Treverton (2001), p. 11, or Nye (1994), pp. 82–93.

are just as secretive as the Soviet Union but more adaptive and faster moving. Typically, puzzles deal less with people and more with things, such as weapon systems, munitions, technologies, and capabilities.

However, for the terrorist threat, not only can intentions not be determined by looking at capabilities, but capabilities themselves have a strong mystery element to them. Much of Soviet capability creation could safely be assumed to be relatively independent of what we did; as one U.S. defense secretary put it about Soviet nuclear programs: "when we build up, they build up; when we slow down; they build up." Not so for terrorist capabilities. Because terrorism is the tactic of the weak, terrorists have to be flexible. They cannot be understood in isolation from what we are doing to counter them. Even their capabilities turn on us. The September 11th hijackers did not come to their tactic as a preference; they chose it because they had found seams in our defenses.

At the end of the mystery-framing continuum would be political and societal questions related to people, such as regional issues, national intent, or group intentions and plans. Understanding human behavior is much more a matter of subjective judgment, intrinsically less certain, than solving problems of science and engineering. The logic train is different for mysteries because no data can "solve" them definitively. They can only be framed, not solved, and thus the logic of argument and analysis is as important as the evidence, often more so. Both types of challenges—puzzles and mysteries—benefit from deep understanding and talented specialists. In both cases, analysis benefits from a blend of expertise in facts with an aptitude for problem-solving, abilities that may or may not be present in the same person.

Analytic Skill Sets and Major Tradeoffs

For both kinds of analytic challenge, progress entails combining analysts who know a great deal about some or many things (experts) with those individuals who can pull together many sources of data and information and, using that evidence, identify new patterns or trends and develop an understanding and new knowledge about complex subjects (analysts). Mysteries involving human perceptions and collective group judgments within a given culture, or between multiple cultures, can also benefit from the insights of intelligence analysts who are not skilled at solving puzzles. Rather, these individuals are skilled in the analysis "tradecraft" of dealing with human behavior within complex political, religious, and cultural contexts. Typically, these people have learned through the experience of dealing with intelligence mysteries over a long period of time.

For analytic groups that regard themselves as being in the mysteries business, such as INR, substantive expertise is more important than puzzle-solving skill. Regional and national problems related to human intent will typically benefit from being addressed by experts immersed in the attendant culture and language, but even then the challenges are formidable. For groups that typically deal in puzzles as much as mysteries, such as the CIA's DI, the two types of analysts are also important but, again, tradecraft focused on culling wheat from chaff, quickly, is essential in dealing with very new political circumstances or subjects. One major dilemma in this regard is that if the situations are truly new, historical patterns may be irrelevant, and so, too, the attendant arithmetic of searching for them. Furthermore, it is unclear almost by definition when a pattern is "new."

Other factors range from deciding how good the Intelligence Community is at capturing a range of views to whether it has the right balance between in-house and external subject matter experts. To the extent that question-answering is the task, it is difficult to query outsiders quickly enough to address highly time-urgent issues, since they will have their own demands to deal with (usually first). Indeed, the very term "outreach" does connote something additional, not a fundamental part of the job. Journalists are often more adept at using outside expertise, since they frequently are able to archive footage in anticipation of newsworthy items—a strategy the Intelligence Community simply does not employ.

The vast majority of intelligence analysts reside outside the Central Intelligence Agency and do work that is tactical, operational, and current. The exact number of analysts is classified but most citizens would be surprised to learn that there are as many analysts in the Directorate of Intelligence of the DIA as in that of the CIA, or that the NSA and the NGA each has several times as many analysts as the CIA.

The focus on the current and tactical intelligence needs of today is pervasive. Even at the CIA, the premier publication has been the PDB. (Responsibility for the PDB has moved to the DNI, but the bulk of the items in it come from the CIA.) Short pieces are favored for the PDB, and these are often based on new information from a secret intelligence source, human or technical. Those secret sources represent intelligence's principal uniqueness and comparative advantage, and they are analysts' focus. Accordingly, career rewards frequently follow short-turnaround reporting, not deep analysis of a particular subject.

If the balance has shifted further toward current intelligence in the last decade, that shift has occurred for several reasons. The Cold War concentration on the Soviet Union and its rather predictable military acquisition system created an Intelligence Community cadre of experts on Soviet military affairs and government activities. Technical issues associated with Soviet weapon systems, defense R&D, and testing anchored a generation of experts on Soviet expertise in functional and operational technical areas. With the end of the Soviet Union, the Intelligence Community was left searching for new missions and new customers, and the deep expertise was assigned to non-Soviet issues if they were available or the experts simply left the service.

The Intelligence Community found new customers in such places as the Department of Commerce, for example, which was more than happy with current reporting. At the same time, though, the end of the Cold War brought declining budgets, leaving the analytic community stretched to serve more customers with fewer resources. As a service business, intelligence has had (and will always have) a difficult time saying "no" or ending service to particular categories of consumers. The analytic community today is the product of the 1990s and reflects a pervasive uncertainty about the emerging, future world.

The portrayal of national intelligence requirements can add to this problem. As noted above, National Security Policy Directive (NSPD)-26, for example, established an NIPF for setting priorities among tough intelligence challenges over the long term. Critical topic areas and challenges are organized into groupings, called bands, running from A to C, with A being a group of the most pressing and important challenges for the nation. Both nations and non-state actors are then extensively listed and their importance in relation to each banded topic areas evaluated and numerically weighted—a factor labeled "propensity." A national intelligence priority scale is then established as the product of the two (band position times propensity). The highest intelligence priorities are then A-band issues associated with nations or nonstate actors with the greatest propensity to engage in that issue—to the potential detriment of the United States. This NIPF has great value for many uses, but it also provides an incentive to reduce spending resources on all but the hottest current priorities, often at the expense of deeper assessments of longer-term challenges.

The Intelligence Community's personnel practices also contribute to this problem by encouraging the creation of generalists rather than specialists. (The future of compensation reform, and the risk that it might exacerbate this problem, are topics addressed in a later chapter.) Unlike some intelligence counterparts abroad, such as the British, who encourage if not reward expertise, most U.S. analysts are encouraged to keep broadening, not narrowing, their focus through a decades-long career. Moreover, the creation of thematic intelligence "centers," such as the Counterterrrorism Center (CTC) or the National Counterterrorism Center (NCTC), also tend to give pride of place to the hottest issues, dedicating only limited resources to longer-term analytic challenges. Global coverage also may suffer in centers that are totally dedicated to themes and therefore may focus on a few "hot" countries. The CIA's Office for Weapons Intelligence, Proliferation and Arms Control (WINPAC) was preoccupied with a half dozen or so countries of particular concern.

The other reason for the dominance of reporting is the by-product of a great success. As technology improved, the take from the big national collection systems for signals and imagery (SIGINT and IMINT) became more available for warfighters. Earlier, when retrieval times were days or weeks, such intelligence was useful mostly for analysts seeking to solve long-term puzzles about Soviet capabilities. Once it could be available in hours, now minutes, in some specialized cases, it became useful to warfighters on the battlefield, and that task came to dominate.

This was the shift to "support to military operations," or SMO. The National Imagery and Mapping Agency (the previous name for NGA) was created as a "combat support agency," working as much or more for the Pentagon as for the DCI and the Intelligence Community. Most of the analysts, especially at NSA and NGA, are in the "force protection" business today, providing tactical support to protect U.S. forces and destroy enemy ones. A congressionally mandated review of NIMA, which reported nine months before September 11th, was eloquent on the risks of the shift: "The need to precisely engage . . . any and every tactical target, without collateral damage, without risk to American lives, requires exquisite knowledge immediately prior to, and immediately subsequent to, any strike. Demonstrably, US imagery intelligence cannot support this activity on any meaningful scale without precarious neglect of essential, longer-range issues. . . ."[3]

At one level, there is broad agreement that analysis should be organized more around issues or problems and less around organizations or collection sources. Yet that proposition remains an issue for the future, as is discussed below. The big analytic organizations, such as the CIA's DI, are more than fully occupied providing a stream of current reporting and analysis. They worry that more "center-itis"—that is, the creation of more problem-oriented analytic centers, real or virtual—will dilute their capacity, and they resist thinking of themselves as being like the military services—that is, as "Title 10" recruiters, trainers, and providers of analytic labor for issue-oriented centers.[4]

At the same time, INR and the CIA's Foreign Broadcast Information Service (FBIS), now the DNI's Open Source Center (OSC), are and feel they must remain in the global coverage business.[5] For OSC, that is so because other agencies are less and less in that business. For INR, it is so because the Department of State is organized regionally and is global, since characterizing and engaging countries is its job, so INR must mirror both attributes.

A related proposition—that analysis ought to drive collection, not vice versa—also commands broad agreement at the level of principle. Yet how to make it happen is also less clear. Postwar intelligence history is littered with attempts to induce policymakers to work with analysts in driving collection. There surely does seem a need for more analysis of targets at the front end of the intelligence cycle portrayed in Figure 2.1. But there is also skepticism, in both the analysis and collection communities, that analysts can know enough about collection system details to make it worth spending much time on formulating a list of intelligence targets for collection and translating them into a collection execution plan.

For the big data processors, NGA and NSA, the challenge is to move from processes that are driven purely by the data collected to ones driven by the problem to be solved. To use a popular but perhaps misleading phrase, it is moving from analysis based on what has been gathered (that is, collection-based analysis) to analysis based on what needs to be hunted (that is, problem-based analysis). Better said, the future will require structuring analytic organizations to do both. For NSA, the Rebuilding Analysis (ReBA) initiative seeks to use analysis to better drive collection as well as produce a higher order of analytic product.[6] As part of this move, the NSA collection and analytic communities will be more proactive in offering targeting support to military consumers, actively engaging in investigations and forensic analysis, and, especially, trying to forecast future SIGINT targeting and development needs.

---

[3] Report of the Independent Commission on the National Imagery and Mapping Agency (2001); the quotation is from the executive summary.

[4] The national 9/11 panel recommended issue-oriented centers as the organizing principle for U.S. intelligence: National Commission on Terrorist Attacks Upon the United States (2004). The specific recommendations are summarized in the Executive Summary and spelled out in more detail in Chapter 13, "How to Do It? A Different Way of Organizing the Government."

[5] This agency is the U.S. intelligence community's primary collector of foreign open information on matters of national security. The unique resources it provides are available to U.S. government and government contractors through a password-protected, state-of-the-art website.

[6] See, for instance, Percivall and Moeller (2004).

For both NSA and NGA, the move toward more problem-driven collection raises questions about how different the styles of analysis are and thus how different the requirements for analysts might be. For example, success in the collection, processing, and rapid exploitation of signals and imagery has spawned a new distinction between fusion—information co-registered to a precise geographic grid, usually under the pressure of ongoing operations—and traditional, longer-term analysis. NGA's concept of "geospatial intelligence" and its fielding of a geospatial framework for use by analysts provide a much richer baseline from which to conduct analysis, whether that analysis is operational or strategic in nature.[7] Building and maintaining the framework is primarily "gathering," which requires a highly efficient production process. By contrast, "hunting," problem-centric analysis, requires empowering analysts in unfamiliar ways, ones very different (and in some cases in opposition to) the familiar production processes.

Examining the Interviews in More Detail

Our interviewees had very different views about the state of their craft. Some analytic agencies report that "all is well" but do so in several variants. Variant 1 is "all is well here but not elsewhere." Variant 2 is "all would be well here if we just had more resources." Variant 3 is that other analytic agencies indicate that all is not well, in the respondent's own agency or elsewhere.

How do analysts feel about their trade? The most frequent lament among all-source analysts relates to data-sharing and data ownership. However, another common concern is the one mentioned above: The Intelligence Community used to do analysis but mostly now does reporting. For many, that is a real lament. For others, such as the CIA's DI, the good news is that consumers, including the president, want their work and, by some judgments, that analysis (or reporting) is more sophisticated than it has ever been.

By contrast, the perception of intelligence, analysis included, in the minds of critics, many legislators, and perhaps the interested public also is that not all is well, far from it. The litany of intelligence "failures," by no means exhausted in Table 2.1, indicates a very different view than that suggested in most of our interviews.[8] By some lights, both the number and the rate of analytic failures seem to be accelerating, and the consequence of them may be increasing dramatically.

To be sure, intelligence analysis always has been as difficult as any human endeavor. Successes often go unheralded, whereas the errors become near-instant banners of disapproval, all the more so now when administrations are more tempted to use intelligence, publicly, in support of their chosen policies. However, if the analytic community is to continue to add value to the choices made by our national and military leaders, it will require a highly creative and highly effective combination of people and technology and an understanding of how

---

[7] For background on NGA goals and framework, see Dervarics (2005). On the challenge of geospatial collection and analysis, see: "Tasking, Processing, Exploitation & Dissemination (TPED) TPED Analysis Process (TAP)."

[8] This list was compiled from a variety of newspaper and other public sources.

Table 2.1
Selected Examples of Intelligence Lapses

1940s	U.S. intelligence predicts that the Soviet Union is five to 10 years away from developing a nuclear weapon. The Soviets detonate a test weapon the following year (1948–1949)
1950s	Intelligence reports warn of a Soviet lead over the United States in missiles and bombers. The first spy satellites put in orbit, beginning in 1960, find no such disparities
1960s	An intelligence estimate says that Soviets are unlikely to position nuclear weapons in Cuba. CIA Director John McCone disagrees and orders more surveillance flights, which soon find signs of missile deployment. Cuban leader Fidel Castro is forced to remove the missiles after President Kennedy orders a U.S. naval blockade of the island (1962)
1970s	Persistent shortfalls in estimates of Soviet military capability and expenditure spark "Team B" challenge to the CIA
1980s	U.S. intelligence fails to predict the impending collapse of the Soviet Union
1990s	United Nations (UN) inspectors discover an Iraqi nuclear program that was much more extensive than the CIA had estimated (1991)
	India and Pakistan conduct nuclear tests . This testing was not predicted by the CIA (1998)
	U.S. warplanes accidentally bomb the Chinese Embassy in Belgrade as a result of erroneous target information provided by the CIA. Three Chinese journalists are killed (1999)
	Significant overestimate of the foreign consequences of Y2K issues (1999)
2000s	The CIA fails to forecast 9/11 attacks. It tracks suspected al Qaeda members in Malaysia months before but fails to place Khalid Al-Midhar (one of the 9/11 hijackers) on its terrorist "watch list" (2001)
	Iraqi WMD estimate took 20 months to develop and was dead wrong in its assessment of Iraqi WMD.

NOTE: In 1976, George Bush commissioned outsiders, a Team B, to critique the findings of the Intelligence Community.

they interact. It will also need, and need to communicate, a more explicit understanding of the limits of analysis under varying conditions and circumstances. Consumers often want definitive answers—point predictions—but intelligence can rarely provide them and should seldom pretend to try.

Although analysts and their managers work in the current glare of public concern over their craft, they have specific concerns about that craft as well. Drawn from more than several dozen discussions and interviews with working analysts and managers, the issues raised below represent those of most concern. And although we conducted too few interviews overall with too many agencies for responses to be statistically significant, they illustrate the concerns raised by a wide range of people involved in analysis across the National Foreign Intelligence Program.

The percentages shown in Table 2.2 simply record how many respondents identified a particular issue as of concern in their interviews. If all 37 interviewees identified an issue, then that issue would receive a score of 100 percent. But the analysis is frequency analysis, not polling; the issues were not set up as mutually exclusive, so responses will not sum to 100 percent. The critical inferences, then, from this analysis are the relative response rates among the issues. For instance, issues related to "Tools of Intelligence Analysis" were generally thought to be

Table 2.2
Analytic Concerns, by Frequency of Mention

	What is (the unique value of) the national intelligence endeavor?	30%
1	Need to redefine intelligence in the future world	30%
1a	What is unique about intelligence? (most often compared to CNN, media)	22%
1b	Unclear goals/objectives prevail	8%
1c	What is newsworthy is not necessarily valuable . . . or intelligence	3%
	Leadership	11%
1	Leadership must be improved . . . it is lacking and problematic	11%
2	Radical transformation of the Intellience Community is needed	3%
	Evaluation	38%
1	Evaluation is critical and needs to be improved	16%
2	Performance (assessment) should be product-driven not "production-" or quantity-driven	14%
3	The PDB is problematic as a performance metric or incentive	22%
4	Evaluation is not critical/desirable	5%
4a	Metrics are not needed	3%
4b	Evaluation is negative	3%
4c	Audits are difficult	3%
	Quality of intelligence	54%
1	Quality of analysis is a concern	14%
2	Focus of intelligence is too "narrow" (e.g., driven by DoD intelligence needs); broader coverage and capacity are needed	14%
3	Analysis is overly current issue demand-driven, needs more long-term view	30%
3a	The long- vs. short-term focus problem is overstated	3%
4	Depth vs. breadth tradeoff must be done better (conflicting demands diminish capabilities)	5%
5	Approach analysis in "problem-centric" manner (vs. geographical) to improve intelligence	11%
6	Global coverage is too difficult; should not be goal	3%
6a	Global coverage Is important	11%
7	Greater client contextualization is needed in intelligence products	5%
	Targeting analysis	32%
1	Targeting analysis is important	30%
1a	Targeting analysis needs to be prioritized/integrated into collection	27%
1b	Targeting analysis is critical and needs more attention	8%
2	Targeting analysis is not needed—leave it to the collectors	3%
	Collection issues	41%
1	Collection strategies and targeting models are dated	8%
2	Law enforcement/DHS agencies hold high potential for collection	5%
3	Science and Technology (S&T) analysis is underused and needs to be better understood	19%
4	Open-source intelligence is critical and needs to exploit new "sources"	11%
5	All-source intellence is critical and materially lacking	3%
6	Need to guard against "evidence addiction" (preoccupation with evidentiary collection)	3%
7	Need to take into account "feedback"—what is the effect of intelligence-gathering on target behavior . . . including denial and deception	3%

Table 2.2—continued

	Tools of intelligence/analysis	54%
1	"Tools" of analysis are inadequate	22%
2	Development of "appropriate" technology is important and directed R&D funds are needed	30%
3	"Tools" limit analysis and are limited by culture/policy/cognition, etc., not just technology	22%
	Staffing	43%
1	Analysis training and education is important and not sufficient or consistent throughout the Intelligence Community	27%
2	Concern over lack of staff/future staffing/"surge" capabilities	8%
3	Analytic corps is highly trained and analysis is better than ever	5%
4	An Intelligence Community–wide curriculum is desirable	8%
4a	Should train stovepipe analysts not Intelligence Community analysts	3%
5	Language and cultural and regional understanding are significant weaknesses	14%
5a	Language and culture knowledge are a strength	3%
6	Career track needs builiding	5%
	Intra-community collaboration/data-sharing	43%
1	Lack of leadership and critical mass impair improving community-wide coordination	11%
2	Stovepiping is problematic and cross-community collaboration is needed	5%
3	Should purse a "virtual organization and fully wired digital network"	
3a	Yes	5%
3b	No	8%
4	Information technology infrastructure, community-wide, is needed	11%
5	Security/secrecy is a concern for virtual organizations and better collaboration	8%
6	Nontraditional source agencies need more input to proceess (DoE Labs, federally funded research and development centers (FFRDCs)	11%

more of a concern by respondents than "Collection Issues." For an initial, quick analysis, this information is at least a guide to what issues were most on interviewees' minds.

Not surprisingly, at the top of this list are the need to redefine intelligence and the need to recognize its twin, targeting analysis, as a unique form of analysis with unique inputs, tools, tradecraft, and training. Interviewees indicated more specific concerns about targeting, asking to what extent it really drives current intelligence, or why it seemingly gets so little recognition within more customary metrics, such as contributions to the President's Daily Brief. The need to develop appropriate technology—not just technology or tools per se—also elicited a high number of mentions, a notable fact despite the potential that our mission may have suggested to our interviewees what we wanted or expected to hear. Next in the list was the concern that the training of analysts is too little and too inconsistent across the Community.

CHAPTER THREE
Key Themes for Leveraging Future R&D Priorities

Intelligence analysis depends on the quality of its people, first, and, on the tools with which they work, second. This chapter focuses on the second, and Chapter Four focuses on the first. An aggressive and focused R&D program can improve analytic tradecraft in the future in many ways. Planned analysis experiments and demonstration tests that compare and contrast different analytic methodologies; well structured table-top games; better time-management tools; advanced, user-friendly analysis software tools; and improved mechanisms to take advantage of relevant R&D outside the NFIP will all improve the performance of future analysts.

A Pyramid of Analytic Tasks

Analysis is definitely plural, a point underscored by Figure 3.1, which also summarizes the key themes that emerge from looking at the analytic community of today and tomorrow. As we have noted—and all seem to agree—the general rubric of "analysis" covers a broad range of activities, each involving its own special set of skills and analytical tools. The figure shows a notional hierarchy of types of analysis beginning with the initial processing of raw intelligence data and extending all the way to the kind of coherent synthesis of a problem that policymakers need to make decisions. Along the way, there are intermediate levels of analysis that are useful to different sets of customers in their own right, in addition to being building blocks for higher levels of analysis.

However, as we have stressed and as the figure notionally indicates, analysis is rarely this tidy. Occasionally, information from the lowest level can feed directly into the highest-level policy documents, and intermediate-level information may also skip echelons from lower levels into higher levels. This compounds the complexity of assuring audit trails for intelligence products.

Ideally, the first stage of analysis involves all the initial manipulation of raw intelligence data that come in from various collection systems to make those data intelligible and useful. As indicated above, in some cases that might be done earlier and presented automatically—computers talking to computers to get target coordinates into the cockpits of pilots as fast as possible. In other cases, that first state of analysis can involve a massive amount of work, some of it quite specialized. Accordingly, special skills and training are necessary for some kinds of

**Figure 3.1
A Pyramid of Analytic Tasks**

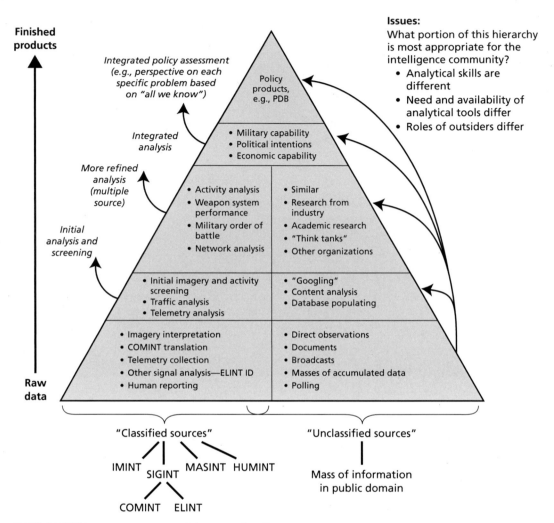

NOTE: MASINT = measurement and signature intelligence.
RAND *TR293-3.1*

tasks. And special analytical tools are extremely useful in some areas—for instance, COMINT initial collection and analysis, imagery analysis, multi- and hyperspectral imagery analysis—to reduce the workload to a manageable level. Since the Community continues to have a problem analyzing all the data collected in a timely manner, there is clearly room for better analytical tools—including operational concepts—in these areas. Some of the analytical skills are relatively unique to the Intelligence Community, but others—telemetry analysis, for example—have industrial analogues and could, in principle, be contracted out. Others, such as language skills, can be acquired without specialized intelligence or industrial training.

It is at this level of analysis where analytical skills are most unique to the Intelligence Community. Although there is obviously some overlap with the "civilian" world, this is where the difference between "secret" information and information from open sources is most readily apparent. The higher one goes in the hierarchy, the more the distinction blurs and the less special intelligence analysis becomes, which is probably why the Intelligence Community has traditionally organized around and focused its analysis on collection systems. This set of special information sources—understood in both a current and historical sense—is the unique contribution that the Intelligence Community brings to the table. Transcending those sources and integrating "secret" information with other data into knowledge that will be of use to policymakers is perhaps the Intelligence Community's primary analytical hurdle.

The next level shown on the chart is a modest refinement of the first, and so the distinctions are necessarily a bit fuzzy. For example, screening imagery for things of interest is critical in military support where time is of the essence. (Past RAND work has analyzed this problem and the associated analytical requirements in some detail.[1]) It is an area where better analytical tools—for instance, automatic target recognition—are essential and may or may not prove feasible. Some of the grander intelligence visions of the future—for instance, global situational awareness—depend critically on quantum leaps in interpreting massive amounts of data rapidly.[2] How well this kind of analysis works is likely to be a major factor in defining the future role of the Intelligence Community, given its traditional emphasis on more and better collection systems.

The middle level of the hierarchy involves more refined analysis of data from multiple sources and integration of that data into useful knowledge. A typical example is defining the characteristics of and developing performance estimates for foreign weapon systems. That means taking the intelligence data, in whatever form and from whatever source, and applying scientific and engineering models to determine the capabilities of aircraft, missiles, and other types of weapon systems. This exemplifies the solving of intelligence puzzles that was central to many of the Cold War debates, such as that over the accuracy of Soviet missiles and thus the vulnerability of U.S. missiles.[3] In those debates, even some high-level policymakers were interested in analysis at this level. In any case, there is a large potential set of customers in industry, middle levels of government, academia, and elsewhere for this kind of information.

Insiders Versus Outsiders

As this example indicates, the Intelligence Community, defined narrowly as the agencies themselves, may not necessarily be the most capable institution for doing this kind of engineering-level analysis. The best people to do it are those who build weapon systems for a living or, depending on the level of detail required, academic specialists or think tank analysts. Even intelligence analysts who invest the necessary time and effort developing suitable analyti-

---

[1] Glenn C. Buchan and others have assembled an unclassified summary of a body of RAND work on that subject..

[2] See, for example, "Global Awareness" (1997).

[3] As Director of Central Intelligence in 1976, George Bush commissioned outsiders, a "Team A," to report on Soviet strategic objectives, missile accuracy, and air defense. The first set of issues became a major political controversy. See Freedman (1986); and Prados (1982).

cal skills to do this analysis will have difficulty staying current, since they are not routinely immersed in it. Parallel examples exist in areas where unique cultural, sociological, econometric, scientific, or other knowledge skills exist. Several programs discussed below—one by the National Intelligence Council and one within the S&T domain—seek explicitly to foster links to outside expertise. The Intelligence Community has solved this problem in the past by contracting with industrial organizations that do this kind of work.

That may still be the best approach to solving the problem, but it raises two issues. One is the question of what kind of skills the Intelligence Community "analyst" needs. Do intelligence agency officers become contract monitors? An analyst who is to be primarily a contract monitor need not have all the analytical skills of one who does the actual work but still needs to know enough to be an intelligent consumer and to integrate the contracted work into the flow of intelligence.

The other issue is whether there will be capacity in the civilian sector. Our interviews, especially those at the National Air and Space Intelligence Center, suggested that the potential contractor base has shrunk significantly, so relying on it may not be a practical option in the future. If intelligence agencies must then do this kind of analysis in-house, even acquiring and learning how to use the standard analytical models will require a substantial investment in manpower.

In other areas, the problems may be more tractable for the Intelligence Community. For example, assessing military orders of battle using intelligence data requires neither elegant analytical models nor very specialized expertise. Since being close to the data helps, the Intelligence Community is probably the logical group to do that kind of analysis. Similarly, the careful work of defining organizational relationships, identifying key individuals, and identifying networks (e.g., terrorist networks) requires classical intelligence analysis methods. Effective computer networks and database management can facilitate the work but they do not change the nature of the basic tasks. Thus, at this level of analysis, several distinctions begin to emerge:

- For some kinds of tasks, the Intelligence Community is probably the most logical group to do the work, and it can probably be done in-house;
- For the most part, this kind of analysis does not require new analytical tools, although more effective methods for creating, exploiting, and sharing databases could make it easier for analysts to do their jobs; but
- Outsiders, perhaps under contract to the Intelligence Community, can do other tasks more effectively. Trying to do the work in-house would require a considerable investment in training of analysts and acquiring and maintaining suitable analytical models. Even if the community were to make that investment—perhaps because of a shrinking base of suitable outside contractors—it would be difficult for intelligence analysts to compete by becoming expert because of the culture and environment. World-class work of this sort does not occur inside the Intelligence Community.

At the next two levels of the analytic hierarchy, the comparative advantage of intelligence may diverge between puzzles and mysteries. For both, that comparative advantage in general derives from two sources. The first is "secrets"—information obtained in special ways that are the unique assets that the Intelligence Community brings to the table. The other is knowledge of and connections to the policy community. Intelligence should know more than outsiders about what is on the agenda, what the frame of the issue is, and thus what should help in the making of policy. For puzzles where secrets matter, intelligence almost by definition has no peers, although in a more transparent world it does have competitors.

Intelligence's franchise at these levels is harder to identify for mysteries. It does have its connections to the world of policy. However, the skills for doing that final integration and synthesis of ideas that policymakers need to inform decisions may be different from the special expertise of the Intelligence Community, which is collecting data and extracting specific kinds of information from it. Performing that integration and synthesis function is likely to require greater understanding of the policymakers, their agendas and predispositions, and the forms of presentation or argument that they will find congenial. It may be a lot to expect for analysts to develop both sets of skills, particularly when the Intelligence Community culture does not really promote or encourage the latter. Thus, the traditional focus of the Intelligence Community on collection rather than analysis may be, somewhat paradoxically, understandable and perhaps even desirable.

Especially in the area of mysteries, America's own analytic history and the culture of intelligence-policy relations may turn out to be handicaps. Harried policy officials often want "the answer"—single-point projections—even about mysteries and even if they know that quest is unreasonable. Intelligence needs to be franker about what it can and cannot provide, defining its assessments as probabilities or ranges of answers.[4] Although this will be less satisfying to decisionmakers, who are pressed to act, a range probably offers a richer portrayal of the actual intelligence, including what is known and not known, and therefore potentially forcing the decisionmaker to develop and address a much wider range of policy options.

As the Intelligence Community is increasingly expected to focus on near-term crisis or "hot-button" issues such as terrorism, its breadth and depth in other areas, perhaps even those of immediate interest, will suffer. In times of tight budgets, it is unlikely that the Intelligence Community will be able to grow enough and develop or maintain the skills to do everything—to deal with short-term critical problems while covering the rest of the world and its various problems in depth. One way to deal with that problem is for the Intelligence Community to evolve into a much larger but distributed and "virtual community"—one that includes a much broader range of topical experts, an idea discussed in more detail below. Another is for the intelligence community to accept a more supporting role, focusing on collecting secret information on selected problems that matter and leaving the synthesis and more extensive analysis of the world to others. Here, the Intelligence Community would become more of a "systems engineer" for a sophisticated set of knowledge and perspectives pertinent to an issue or a threat.

---

[4] Sherman Kent charmingly recalls his effort to add precision to such estimative words as "likely," by defining them in terms of probability bands. He was opposed by two very different camps in the Office of National Estimates: The "scientists" regarded the attempt as spurious precision and the "poets" saw the whole enterprise as too iffy in any case. See Kent (1994).

The analytic community's situation today may bear analogy to the nation's science and technology base at the end of the Cold War. Through the 1980s, the driver for high-tech R&D was the government; defense and intelligence led the way. They had the requirements and the budget and "drove" most of the high-tech R&D innovation. By the 1990s, however, industry (much of it the software, information technologies, and entertainment industries) was in charge, driven by the market and by Moore's law of geometrically expanding information technology. So, the expertise, resources, and "pull" of defense and intelligence attenuated. That trend continues today.

Perhaps somewhat analogously, the "outside world" is just as awash in data as is intelligence, given the Internet, globalization and digitization, multimedia, and the like. Although outside groups are sometimes less technically complex, as are the volumes of information they collect, their need to develop the tools to search, cull data, find trends, and even automate translation has begun to rival the need of the Intelligence Community. As with high technology R&D, the market is the driver.

So intelligence is no longer the "keeper" and sole developer of, nor the only customer for, tools to search, analyze, and synthesize massive amounts of data. Accordingly, just as intelligence has had to increasingly rely on outside S&T sources, it will have both the need and the ability to rely on and cultivate outside sources to support analysis, whether that means software tools or academic experts. There are more of both out there than ever before. What intelligence used to "own," it now can and must "buy" or otherwise tap.

The Range of Tools

Too often "R&D" equals "tools." The thrust of our conversations and our analysis is that the equation is wrong. Indeed, tools may be the least relevant R&D product for the analytic community. Accordingly, this section starts by discussing tools, then it moves to who does what. It should be read in the context of the previous and following sections: Tools are helpful but not essential. What are essential are the people skills—training, policy, and leadership. Moreover, R&D could also be shaped and focused through frequent Community-wide experiments that bring disparate elements of the analytic community together around a common problem in a way that emphasizes data-sharing and alternative approaches to analysis.

The definition of tools in this report is broad but not boundless. Tools here means technologies, products, or processes that will help analysts in three ways—first, in searching for and dealing with data (easier access to databases, improved search engines, better algorithms and ways for looking for and displaying patterns or outliers, and the like), second, in building and testing hypotheses (for instance, economic models or methods, such as the Delphi techniques or factions analysis, for aggregating subjective judgments[5]), and third, in communicating more easily both with those who will help them do their work (from analytic work-

---

[5] For a discussion of a variety of RAND strategic planning methods, especially assumption-based planning, which seeks to identify the critical assumptions on which current policies are based, see Dewar (2002). One of the seminal RAND works was Kahn et al. (1976). A recent exemplar is Lempert, Popper, and Bankes (2003).

ing groups to collectors and collection managers) and with consumers, both to deliver their analyses and to better understand what consumers want. For this report, tools do not include, for example, important improvements in supporting software that will enhance daily work flow efficiencies and work planning and scheduling capabilities. These will be advanced by the commercial sector in ways that will need limited help from the government.

There is no consensus on the need for or value of "tools." In our conversations, views on tools ranged from "the best thing since sliced bread" to "evil and nefarious." Issues include suspicion of new technologies, templates, filters, and profiles because analysts want to control the analysis themselves, with others claiming that a handful of analysts could cover the work of a thousand with pre-organized, preplanned collections, tailored to the original problem-tasking. Analysts often feel that those tools quickly become the province of the software engineers who designed them, and so wind up being unfriendly, if not unusable. One of our interviewees made the specific point that tools generally are interpreted as information technology and just that. Five others noted that their analytic agencies had no budget and no ability to develop tools tailored to their needs. By contrast, two interviewees, both from service intelligence centers, reported themselves awash in tools. As one put it, their analysis is driven not by problems or sources but by tools themselves. Tools define the box they are in, and the tools will not let them out.

Tools have their highest value when they free up the analysts' time to think. There does seem to be agreement that tools can help in only certain kinds of analysis, that the more technically complex they are and the more they require sophisticated inputs, the more difficult they are to integrate into the analysts' environment and to use. Table 3.1 illustrates the range of tools and skills supporting intelligence analyses.

The most obvious problem is that there is no good, Community-wide mechanism to solicit analytic tool "needs" or to establish requirements. Although there are multiple sponsors for technology, in the CIA and beyond it in the wider community, those sources or sponsors are mostly uncoordinated.[6] Nor are there systematic mechanisms for transferring or inserting technology, both among and even within individual agencies. Moreover, and critically, issues about tools merge with issues about the comparative advantage of the analytic community within the government relative to capabilities that exist outside the government and are paced by industry and commercial market needs: What should the Community do itself and what should it outsource? Where are the boundaries of tasks it should undertake at all? Outsourcing is a tool itself that the Community is struggling to learn to use effectively.

Much of the routine development, maintenance, manipulation, and first-order analysis of databases will be facilitated by better computer applications software and systems. A critical issue is developing systems that are sufficiently user-friendly that analysts will be

---

[6] This comment is perhaps less true in the area of information technologies. Not only does the CIA house the Intelligence Technology Innovation Center (ITIC), but In-Q-Tel and its in-house CIA counterpart, the In-Q-Tel Interface Center (QIC), are effectively mapping the needs of the CIA to new technology opportunities. In 2006, the Intelligence Advanced Research Projects Activity (IARPA) was created to provide a focal point for technology development in the Intelligence Community, subsuming both ITIC and the Advanced Research and Development Activity (ARDA).

Table 3.1
Wide Range of Analytical Tools and Skills Required

Stage of Analysis	Example Functions	Types of Tools	Types of Skills
Collection target planning analysis	Translate policy/ military questions into information needs Translate information needs into collections goals Translate collection goals into target lists	Collection system simulations HUMINT planning tools Event correlation and analysis Denial and deception gaming and simulations	Experience in analysis and collection limitations Policy and military operations expertise HUMINT operations expertise
Data processing and exploitation analysis	Raw data processing Imagery interpretation Telemetry initial analysis and identification ELINT identification COMINT translation and screening HUMINT screening	Need for tools to screen, filter, and interpret Standard telemetry analysis software Signal identification and location software Decryption, sorting, and translation software Database-mining solftware, perhaps biometric data analysis	Special skills for different types of imagery (e.g., visual, infrared,radar, hyperspectral) Standard engineering skills Language skills Cryptology skills Human skills that operatives need—insights
Selected source analysis	Imager analysis Telemetry analysis and interpretation Traffic analysis HUMINT interpretation	Specialized tools for each kind of imagery Other more standard tools for "bookkeeping" Data-mining and pattern recognition	Some unique intelligence skills (e.g., "cratology") Uniue imagery interpretation skills Human analysis skills
All-source fusion analysis	Integrated analysis Military capabilities Economic capabilities Political intentions Activity analysis Network analysis Horizontal integration	Global analysis tradecraft Campaign analysis Standard economic modeling System performance models Complex network simulation Interactive research tools	Tradecraft experience Systems, analysis skills Technical and operational insight and syehtsis for military problems Political and economic insight for other issues

willing to invest the time and effort necessary to learn to use them. The analytic community needs Community-wide standardization, not boutique solutions. However, analytical tools, particularly software tools, are critical in some areas, such as technical data processing and when large-scale databases must be evaluated, correlations identified, and false alarms reduced. Some of these tools may require breaking new ground as the amounts of data increase and important trends and patterns become subtler.[7] In other cases, the tools are well understood but require some effort and investment to use correctly. In those cases, the Intelligence Community will have to decide whether developing and maintaining the tools and learning to use them effectively are worth the resources. If the answer is to involve more outsourcing, then the community would be able to focus more on the fusion of these relevant topical analyses into a coherent whole product tailored to both pressing and long-term national and shorter-term military issues.

Since many major analysis organizations (for instance, the CIA's DI and DIA) have no organic R&D budget, they nudge, cajole, and prod the potential developers of tools to respond to their needs. As far as the CIA is concerned, these R&D efforts are concentrated in the DS&T's office of Advanced Technology Programs (ATP) where basic R&D is done. In-Q-Tel and QIC also work to meet CIA users' needs with technology, as does the ITIC for some longer-term needs. ATP inherited a major share of former ORD (Office of Research and Development) staff following that agency's dissolution in 1997. Many of these offices coordinate with other government innovation and R&D efforts, such as those at the Pentagon's Defense Advanced Research Projects Agency (DARPA) or the Intelligence Community's ARDA or the ITIC. This process certainly has merit, but its technical and resource efficacy in dealing with fundamentally different future threats whose characteristics may not be known or easily articulated remains uncertain.

Beyond the resource question is the practical issue that came up again and again in our interviews: Where is the bridge between the developers and users and how is it used? In-Q-Tel's annual classified "problem set" is based on canvassing the potential user community for needs that can be filled with unclassified R&D. It is impressive but is mostly a generic, catchall list, reflecting principally CIA wishes.[8] Although it has moved away from the heavy original emphasis on information technologies, the roster of items does not always address many of the Intelligence Community's most pressing analytical problems.[9] Originally, information technology projects for the Directorate of Intelligence accounted for about 40–50 percent of its budget, but the problem set now includes more infrastructure-related activities and esoteric projects in support of collection and operations. ATP focuses a great deal on classified S&T requirements, with a smaller amount of their work aimed at R&D for users outside the S&T community (e.g., machine translation for the CTC).

---

[7] The ill-fated Pentagon program, Total (later Terrorism) Information Awareness (TIA) proposed "revolutionary technology for ultra-large all-source information repositories," which would contain information from multiple sources to create a "virtual, centralized, grand database." Although it was only a pilot research program, it generated enormous controversy after the *New York Times* reported on it in November 2002 and was formally closed. For background and a range of cites on the program, see http://www.epic.org/privacy/profiling/tia/.

[8] In-Q-Tel's areas of focus are knowledge management, security and privacy, search and discovery, distributed data services, and geospatial information services. See http://www.in-q-tel.com/about/model.html.

[9] This situation has improved markedly as QIC gains more experience and resources.

There is no shortage of relevant technology (not simply software tools) being developed across the Community. It is not difficult to pick a few samples of technology that suggest the range of what is possible in pushing the state of the art in areas related to intelligence analysis. For example, DARPA's EELD program (Evidence Extraction and Link Discovery) was the descendant of a kindred system produced in the late 1980s for the predecessor of the CTC.[10] The goal of the EELD program was to develop technologies, methodologies, and tools for automated discovery, extraction, and linking of sparse evidence contained in large amounts of classified and unclassified data sources. EELD developed detection capabilities to extract relevant data and relationships about people, organizations, and activities from message traffic and open-source data.

It linked items relating potential terrorist groups or scenarios and learned the behavioral and activity patterns of different groups or scenarios to identify new organizations or emerging threats. To illustrate by analogy, the goal was to build on significant technical successes in the signal processing methods historically used in nonacoustic antisubmarine warfare (ASW) to automate an assessment of patterns of human behavior in ways that allow "normal" behavior to be characterized. Once this is done, deviations from normal behavior, even extremely subtle ones, may be detectable and used as a basis for further, more specific investigation that could provide a systematic early alert of threat activity. This has obvious implications for intelligence analysis and may be extremely promising in counterterrorism applications.

Another program, in a second, more advanced phase, was GENOA-II.[11] It focused on developing information technology needed by teams of intelligence analysts and operations and policy personnel in attempting to anticipate and preempt terrorist threats to U.S. interests. Genoa II's goal was to make such teams faster, smarter, and more coopertive in their day-to-day operations. Genoa II tried to automate team processes so that more information could be exploited, more hypotheses created and examined, more models built and populated with evidence, and in the larger sense, more crises dealt with simultaneously.

Genoa II attempted to develop and deploy:

1. Cognitive aids that allow humans and machines to "think together" in real-time about complicated problems;
2. Means to overcome the biases and limitations of the human cognitive system;
3. "Cognitive amplifiers" that help teams of people rapidly and fully comprehend complicated and uncertain situations; and
4. Ways to rapidly and seamlessly cut across and complement existing stove-piped hierarchical organizational structures by creating dynamic, adaptable, peer-to-peer collaborative networks.

---

[10] EELD programs are a continuing theme within ARDA. For background on EELD, see http://www.rl.af.mil/tech/programs/eeld/.

[11] See http://www.eff.org/Privacy/TIA/genoaII.php and http://www.darpa.mil/DARPATech2002/presentations/iao_pdf/slides/ArmourIAO.pdf.

These programs are formative and are driven by the recent emergence of technologies that may actually allow them to be real. However, those technologies are rooted in principles that have yet to be tested in operational settings. For example, most such programs presumed that collaboration among peers is good and will lead to better decisions. Yet the conditions under which this is true, or not, have yet to be clarified.[12] Again, the Community-wide potential value is apparent but is not being evaluated and an intimate development and testing partnership across the Intelligence Community does not exist.

Other related programs existed in other agencies. For example, one sponsored by ARDA and NGA was NIMD (Novel Intelligence from Massive Data).[13] It aimed at focusing imagery and geospatial analysis attention on the most critical information found within extremely large datasets—information that might indicate the potential for strategic surprise. Novel intelligence was defined as actionable information not previously known to the analyst or policymaker. It may give the analyst new insight into a previously unappreciated or misunderstood threat. Massive data may be "massive" because of the sheer quantity of similar items. Or a smaller volume of data may nonetheless be considered "massive" because it consists numerous types and formats of data from separate sources—structured text in various formats, unstructured text, spoken text, audio, video, tables, graphs, diagrams, images, maps, equations, chemical formulas, and so on. Data may also be deemed "massive" because of its inherent complexity, which arises when a single document contains links between multiple information sources, with the meaning of any source dependent on information contained within other sources.

Understanding the content of complex data requires the ability to process data that has already been combined from multiple sources—a task beyond the capability of current technology. A deeper level of complexity comes into play when information requires various kinds of expertise. For example, analysts might need to consider social, military, economic, political, governmental, scientific, and technical issues surrounding an event or location. The NIMD problem is made harder by the realities of how the human mind works. For example, the order in which people read documents has a great deal to do with the hypotheses they develop and the weight they assign to further readings. They also tend to "anchor" to one favorite hypothesis very early in the process of information exploration, discounting content that does not support the pet hypothesis and overestimating the worth of documents that do.

Thus, NIMD was about human interaction with information in a way that permits intelligence analysts to potentially spot the telltale signs of surprise in massive data sources— building tools that capitalize on human strengths and compensate for human weaknesses to enhance and extend analytic capabilities. For example, people are much better than machines at detecting patterns in a visual scene, whereas machines are better at manipulating streams of

---

[12] For instance, although most research suggests that groups can be more creative than individuals working alone, the perils of "groupthink" or "organizational lock-in" on one hypothesis have long been concerns. See, for instance, the classic work, Janis (1972); and van der Heijden (2002), pp. 50–51.

[13] See http://www.ic-arda.org/Novel_Intelligence/.

numbers.[14] NIMD captured and recorded activities that occurred during the course of analysis and could provide an audit trail of how, when, and why analysis decisions were made. At a minimum, this might simplify lessons learned during post mortem assessments.

The basic point to underscore here is that there has been no shortage of efforts afoot to develop tools. What is less clear is whether there is always a bridge to ensure that they find a home. In the case of NIMD, the sponsor(s) required that bidders demonstrate that they had relationships with working analysts and sufficient access to them to develop the technology in concert with potential end users. Some DARPA programs are now requiring this as well. Moreover, both NIMD and some DARPA programs no longer go through the "demo, prototype" phase, but, rather, expect initial proposals to have a technology transfer plan (and experience) to move the work from the laboratory to the analyst's desktop. The success of these bridging efforts has, to date, been mixed. For example, complaints from DARPA project managers sometimes are the obverse of those from the analysts: that the CIA sends researchers over to help form requirements but then the CIA folks disappear or participate only sporadically.

However, there have been successful multiagency development programs. TIPSTER—a cluster of joint projects among government, industry, and academia aimed at improving capabilities to process text—included one multilingual data extraction program that started more than a decade ago.[15] It was so successful it was commercialized under the auspices of the National Institute of Standards and Technology (NIST), having started out in the shadows of the Department of Defense and the CIA. The overall accuracy of the highest-performing TIPSTER text program on news stories is 96 percent on relatively simple tasks, 80 percent on the task of merging various types of information, and 56 percent on the difficult task of identifying events of interest in a given text and merging various pieces of information about each event into a single output.

The development process for TIPSTER was equally impressive. DARPA, the DoD, and the CIA jointly funded and managed the program, in close collaboration with NIST and Naval Command Control Ocean Surveillance Center (NCCOSC). A TIPSTER advisory board was formed with members representing users from other government agencies including the Department of Energy (DoE), FBI, Internal Revenue Service (IRS), National Science Foundation (NSF), and Treasury Department. In all, 19 industry and academic partners were involved in TIPSTER.

Initiatives to Link Tools to Tasks

The Intelligence Community has a long history of commissioning the development of knowledge engineering and knowledge discovery techniques to address the issue of critical analysis and "strategic surprise," but little of this work has seen actual service. To recapitulate the reasons why: First, there is no single point of Community-wide oversight of R&D related to analysis that can assure that the analysts have a full knowledge of emerging opportunities.

[14] Tufte (2001).

[15] For background and a variety of sites on TIPSTER, see http://www.fas.org/irp/program/process/tipster.htm.

Second, no Community-wide mechanisms exist to input analysts' wishes and dreams into the development community. No collaboration paths are in place, Community-wide, to beta-test and evaluate new technology and no well-traveled roads are available to assure that appropriate developments outside the Intelligence Community can be easily integrated into the Community. Although many factors have contributed to this failure, the most chronic difficulties seem always to fall into two categories:

- The techniques fail to bring an experience factor to bear on the problem to acquire or to use the prior knowledge—the "thread of logic"—that analysts bring to their tasks. As a result, discoveries made by machines prove to be trivial, well-known, irrelevant, implausible, or logically inexplicable; and
- The newly developed techniques fail to respect analytic work habits or try to solve tough problems with technical approaches that simply do not work, as was the case in NSA's Trailblazer Program.[16] Many analysts have neither the production incentive nor the introspective bent to interrupt their workflow to learn algorithms about analytic techniques or the tradecraft unique to their jobs—particularly when those techniques themselves may be undergoing rapid change to suit a new domain or information demand. The watchword for analysts is: If it takes more than 15 minutes to learn, I cannot afford it.

Two more recent initiatives have sought to learn the lessons from this experience. Then-FBIS's collaboration with IBM generated data-mining algorithms uniquely tailored to FBIS's needs, along with new search and correlation technologies that have let the organization achieve a seven-fold increase in output with half the staff.[17] The Web crawlers produced by IBM and tailored to FBIS's needs far surpass Google in efficiency. FBIS found IBM's industrial experience and depth essential, and the collaboration was close enough so that what IBM produced actually met the needs of FBIS.

The Trident workstation was developed through the CIA's Chief Information Officer. It involved CIA and TTIC (now NCTC) analysts in the prototyping phases. In potential, it was impressive; it was when demonstrated to us. It assured a distinct segregation between the data layer, which is viewed as the Holy Grail to be preserved in perpetuity, on one hand, and the software layer, on the other, which can undergo constant changes as new modifications, versions, and products are released. Its starting point was that analysts spend 90-plus percent of their time on research and analysis, production of finished products, and communication and administration. The initial thought was to have all three in one "cockpit," but the developers decided to begin by concentrating on research and analysis. All this is driven by the needs of analysts, especially their need to run through several hundred messages quickly each morning before a staff meeting. Adding tools is computer-intensive and so degrades basic retrieval times.

---

[16] Trailblazer was designed to help the agency sift through and make sense of the torrent of data it collects from cell phone conversations, faxes, e-mails, and a wide variety of other electronic communications around the world. See Lewis et al.

[17] The IBM search procedures are described in Carmel (2004).

The Trident developers looked at a number of areas of requirements identified by analysts working within TTIC and the CIA's DI—geospatial, analytic environment, data visualization, information extraction, intelligent agents, knowledge management of what is already known, machine translation, link and relationship analysis, and search and retrieval. The double-size screen begins with a profile of incoming messages, listing them on one side with the message on other. A suite of tools is, in principle, easily available. One, Inspire, can map messages or key words, producing peaks where there is a lot of similarity, thereby letting analysts get a quick picture of their message traffic. A click gets machine translation of a foreign item. Another gets into the video library, also with translation. A mapping function takes analysts to a place and lets them move in and out of detail. Duplicates are stripped out. An entities function allows the analyst to sort messages by people, places, and things—as apparently some analysts do. Intelligent agents will track, say, shipping, between points with minimal datasets. If contacts are listed, analysts can quickly contact them through instant messages. Link analysis will very quickly produce a rough draft of connections.

There are other links between analytic components, such as the CIA's DI, and non-CIA R&D, such as that done at ARDA or DARPA's former TIA project.[18] ITIC seems, though, to target the majority of its investments on basic scientific research areas with limited ties to intelligence analysis and problems in supporting operations. Again, what is lacking is a systematic process that identifies analysts' needs across the Community, surveys the larger defense and intelligence R&D community for potential efforts that could meet these needs, then follows the innovation through to technology insertion and user adoption. There are isolated instances and individual cases where this is not the case, but no true Community-wide effort to leverage common needs and solutions exists, and such an effort would be very cost effective.

---

[18] Poindexter (2002).

CHAPTER FOUR

Building the Human Capital for the Future

Intelligence is about nothing if not about "out-thinking" the adversary. For all the appropriate emphasis on technologies, methodologies, tools, and infrastructure, people are the Intelligence Community's most precious resource. Whatever the changing paradigm for analysis, analysts remain at the center. Training and professional development of analysts remain a challenge. Learning the detailed intelligence-related skills is hard enough; producing world-class analysts with adequate breadth and depth is more demanding still.

The analytic community faces a clutch of human capital issues, from the immediate effect of compensation reform, to how to develop and train analytic tradecraft, to how to nurture a new and different cohort of analysts. There are now several focal points among specific organizations for analytic tradecraft, notably the CIA University's Kent Center but also NSA's National Cryptologic School (NCS), and the Joint Military Intelligence College (JMIC), as well as training programs within the Service Centers such as the National Air and Space Intelligence Center (NASIC) and the National Ground Intelligence Center (NGIC). But there is no regular process for comparing notes across agencies or for learning lessons about what works. Most of the initiative in training, at least beyond initial introductions for new analysts, rests with individuals; there is little strategic view, within agencies or across them, of what skills are and will be needed. The creation of the National Intelligence University opens the opportunity to change that state of affairs.

Reforming Compensation and Incentives?

The government-wide move to reform compensation by moving it away from seniority and toward performance is sweeping and runs well beyond intelligence. Although those reforms have been postponed for the CIA and other intelligence agencies for the time being, they offer opportunities as well as dangers. The Department of Homeland Security (DHS) acquired considerable freedom from traditional civil service rules at its inception, and the Department of Defense was given that freedom in legislation.[1] There are two essential elements to the

---

[1] For detail on the practices of DHS's Transportation Security Agency, see http://www.gao.gov/new.items/d03190.pdf.

reform—new pay scales and new pay decision processes.[2] Figure 4.1 provides highlights of several aspects of this reform initiative.

Each position or job is to be reviewed and, depending on that job analysis, which includes an assessment of the outside market, a small number of "occupations" will be determined. Each job will be assigned to one of three broad occupational bands. Pay scales are based on a market survey of these occupational bands, which is to be updated regularly. Employees will not lose any salary at the time of conversion to the new system. For example, if an employee is due to receive a within grade increase (WGI), he or she will receive a prorated amount of money in salary at the time of conversion.

In deciding on pay, employees will be grouped into pay pools, generally depending on the assigned work unit, and evaluated by a supervisor in the chain of command. In deciding on pay, the pay pool manager is not the employee's supervisor but reviews input from the employee's supervisor in making pay decisions. This is to be done annually. Analysts are currently very driven by immediate demands for products and "pieces." Their world is fast-paced and dynamic. The situation could easily become "what have you done for me lately?" when reviewed by the pay pool manager a year later. For promotion, the decisionmakers are the promotion panels who also review inputs from the supervisor.

Promotions are based on sustained performance and demonstration of skills at, or exceeding, the next career level. Because of the broad occupational bands, promotions will not occur as often in an employee's career in the future as in the past.

Figure 4.1
Proposed Compensation Reform Process

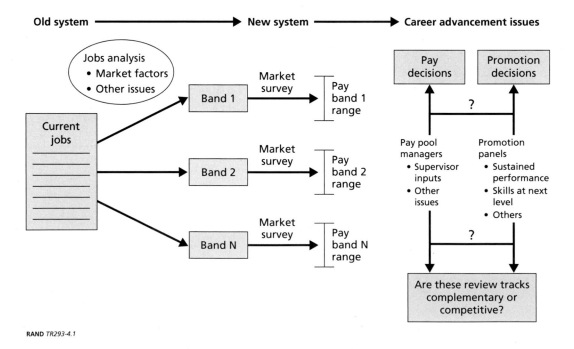

RAND TR293-4.1

---

[2] For background on the reforms at NSA, see http://www.politrix.org/foia/nsa/nsa-reorg-id.htm.

Interviews and, especially, our reviews of dozens of issues of the CIA's *What's News* suggested that many employees see a negative effect on their retirement, and the answers provided by agency leaders fail to assuage their concerns.[3] The concerns include:

- The time that will be spent discussing the new system—occupations, pay decisions, promotions, new review forms, etc;
- How the occupations are determined, including individual assignments to a particular occupation;
- How decisions will be made and who will make them;
- How to have enough background to make investment decisions concerning any bonuses received; and
- How to get questions answered when supervisors are not necessarily familiar with the details of the new system.[4]

Implementation of the compensation reform was pulled back, after first being delayed for the CIA, until FY2005. The 2003 Intelligence Authorization Act included a pilot program to begin in January 2003 and run for one year, to be followed by the BENS assessment mentioned above.[5] In any case, the prospect raises questions for the future of analysis.

Could compensation "reform" further entrench the immediate "publish or perish" culture? Those working on long-term think pieces or obscure accounts might not have visibility and might be less rewarded than they are today. Instead, rewards would go to those visible in, say, producing PDB items or other forms of current reporting. In short, pay for performance is a great opportunity, but much will depend on how it is implemented.

Training Analysts

Plainly, the needs for training vary greatly across the Intelligence Community. For instance, the training for deep subject matter experts would be very different from that of analysts who must be broadly knowledgeable in a range of current priority subjects. The former no doubt would have to bring much of their training with them through prior education. Yet the variations in the training of analysts across the Community are striking—from what was a 20–22

---

[3] The CIA asked Business Executives for National Security (BENS) for an assessment of the CIA pilot program. See Business Executives for National Security (2004).

[4] These concerns are hardly limited to analysts or employees. See, for instance, House oversight committee chairman Porter Goss's comments, in July 2002 discussing HR 4628, The Intelligence Authorization Act for FY2003: "The Administration strongly opposes Section 402, which prohibits the Central Intelligence Agency (CIA) from implementing compensation reform plans. The Director of Central Intelligence (DCI) must have the maximum flexibility in managing the CIA work force to ensure that the Agency can quickly adapt to changing mission demands and personnel needs. CIA's reform proposal is fully consistent with the President's Management Agenda, which aims to pay employees in a way that recognizes their contribution to mission and rewards top performers. At a time when it is needed most, Section 402 would curtail the statutory authority and flexibility that the DCI has had since 1949 with regard to CIA employee compensation."

[5] See Section 402(a)(2) of the Intelligence Authorization Act for fiscal year 2003 (Public Law 107-306; 116 Stat. 2403; 50 U.S.C. 403-4).

week program for the CIA's DI to virtually instant short-course immersions elsewhere. As Rob Johnston, a rare anthropologist who turned his attentions to the Intelligence Community, thanks to the CIA's Center for the Study of Intelligence, has found, most training is on-the-job training.[6] His research dovetails with our interviews in other respects as well. None of the agencies had much familiarity with the analytic techniques of the others. In all, there tended to be a great deal of emphasis on "skill level" certification, organizational processes, and writing and communication skills and much less emphasis on analytic methods. Training was driven more by individual analysts than by any strategic view of the agency or the Community and its needs. There is a striking absence of Community-wide common course components of emphasis on Community-wide perspectives

That driver is one among several that leads toward an emphasis on credentials in training, perhaps at the expense of techniques more directly related to immediate analytic work. The Kent School, for instance, confers bachelor's and master's degrees, and JMIC offers both as well. Surely, there is nothing wrong with degrees or other credentials. However, operators expressed concern that the schools were too distant from the needs of operators to be as helpful as they might be. In the view of the trainers, the concern was keeping up with the pace of needs—usually defined as needs for specific new area or country knowledge—in circumstances where the most knowledgeable possible "teachers" were precisely the experts in highest immediate demand. The Kent School, for instance, offered 80 courses on specialty disciplines.

Initiatives in tradecraft are also isolated. The Kent School, for instance, tried to keep up with best practice in the private sectors by sustaining four small teams. One handled outreach, another focused on product evaluation, a third looked at methods, including tools, and a fourth treated integration, which means trying to keep the school's offerings matched to the needs of the DI. (The Product Evaluation Staff was moved to the DI Front Office.)

We did not survey the curriculums of the various schools in great detail—that would be a very valuable project, one that we suggest undertaking.[7] What seemed clear, though, is that, especially with the rush of immediate need, there are few opportunities or mechanisms for looking at tradecraft jointly, for understanding how other agencies do "analysis" and what might be learned from them, or for developing centers of training excellence that develop comparative advantage instead of duplicating what has been done elsewhere.

Regular joint training experiments and field tests in tradecraft would make sense, first, because there is a basis in analytic methods that is shared across very different tasks of intelligence analysis. Second, that training could begin to foster a greater sense of joint tradecraft—more Community awareness on the part of analysts of what fellow analysts in other agencies do or could do. And the process could contribute to advancing joint efforts more generally across the Intelligence Community. The DDNICAS's creation of a joint Intelligence 101 course is a first step.

The other issue begging for attention is the ratio of outside to inside expertise. Those we interviewed in Air Force intelligence, for instance, feared that they were trading depth for

---

[6] See Johnston (2005).

[7] In a follow-on project, we looked in more detail at the offerings relevant to analytic tradecraft of the various schools. The results of that analysis have not yet been published but are available from the office of the DDNI(A).

breadth, with the situation compounded by a blurring between military and civilian resources and considerable growth in the size of the civilian workforce. This trend was evident throughout our survey. Given the demands of current business, there simply is no time to "train up" an analyst on a new, current, hot-button issue with any serious depth. The National Intelligence Council (NIC) is addressing the issue of outsourced experts through its Global Expertise Resources Program (GERP), which pre-establishes ties to subject matter experts in critical areas throughout the world and facilitates their use to address key intelligence challenges. GERP had a goal of moving toward 100 experts in the future, but other agencies worry that GERP will be too small, or its reservists too far from the needs of policy, to be very helpful.[8] Clearly this will depend on the questions being asked.

A similar program—the Science and Technology Experts Program, or STEP—has been under way for a number of years, although the same concerns have been expressed about its size and the scope of the program relative to demand. As currently configured, the STEP gives the NIC and others access to a few dozen organizations with subject matter experts in key areas of science and technology. This resource may be tapped for specific advice on intelligence problems. However, the resource is primarily useful to efforts of limited scope and short duration. Program activities are manifestly consultative or advisory in nature and are difficult to translate into improved core capabilities for the community.[9]

A New Generation of Analysts

Finally, the next generation of analysts has much more experience with and is much more comfortable than its seniors with information technologies, networked environments, and parallel processing of large amounts of information. These young people access data, share hypotheses, create "problem-centric" networks, and communicate in parallel with their friends in ways that will shape how analysis will be done in the future. Gilman Louie, former president of In-Q-Tel, describes a wide range of technologies and concepts for using them that the modern student uses for purposes of learning, socialization, and accessing and storing data that are a far cry from today's intelligence architecture.[10] The Community will not attract, or will soon lose, these young people if it does not accommodate to how they think and learn.

Now, however, the Community suffers because the tools and technologies are rarely, if ever, available in an open architecture system within the agencies, because of both legacy architecture and the constraints of security. Moreover, that young analysts do not have access to these commonly available tools means that they will have less capability internally (for their job) while having less and less familiarity with the innovative ways that others (in their target community) are experimenting and innovating with the same tools.

---

[8] The program is described at http://www.cia.gov/nic/NIC_associates.html.

[9] In 1994 RAND did a study for the then-Community Management Staff of options for creating a reserve corps for the Intelligence Community.

[10] In-Q-Tel CEO Summit 2003, a collection of talks by CEOs of innovation technology companies.

So far, the analytic community has not given much attention to how new analysts are to be recruited, nurtured, trained, and, importantly, retained. It has shown little evidence that it can effectively use technology advances to improve analytic tradecraft. These raw skills bring an opportunity to the analytic community to leverage technologies and tools in fundamentally different ways, provided these attributes can be effectively shaped and used to enhance analytic tradecraft.

Again, to overstate for effect, the next generation will be fast, not slow; does parallel processing, not serial processing; gives pride of place to graphics, not text; does random accessing, not step-by-step processing; is connected, not stand-alone; is active, not passive; mixes work and play; is impatient for results; mixes fantasy and reality; and very definitely sees technology as a friend, not a foe. These characteristics can be the greatest future assets or considerable liabilities, depending on how these resources can be channeled to solve key intelligence challenges.

CHAPTER FIVE

A Vision of the Analytic Community Tomorrow

What are the boundaries of "intelligence analysis"? Is it primarily the purveying of secret bits, with context? Or is intelligence the integrator of information for policy? This question of doctrine has enormous implications for the analytic community of the future. Although this research certainly cannot provide a definitive answer to this question, it does provides a framework for further debate. This chapter delineates the framework and sharpens the issues that will materially affect the definition of intelligence analysis in the future.

In this light, one core and narrow definition of national intelligence is the product of the endeavor by which politically, economically, diplomatically, and militarily relevant information is determined, gathered, and transformed into new insights and knowledge about all potentially threatening nations, groups, and individuals—information and knowledge that the threat actors do not want the United States to know. However, national intelligence must draw on all information sources, both classified and unclassified, including (necessary but not sufficient) intimate partnerships with open sources, including the media. Finally, the enterprise may or may not distinguish between domestic and foreign intelligence—functionally, organizationally, and legally. Figure 5.1 notionally illustrates the range of issues to be parsed in characterizing national intelligence analysis in the future.

The analysis of most intelligence issues will involve making use of three different types of information in different amounts, depending on the specific issue at hand. The types are information typically widely known about a subject, information that may not be known, even to the relevant parties involved, and information intentionally withheld from the United States. This last category is central to the national foreign intelligence endeavor and underscores its uniqueness in policy formulation activities relative to other organizations, including the media and CNN-like information providers. Given these elements, selected changes in analytic tradecraft, along with future issues, are highlighted below.

Changing Tradecraft Characteristics

Table 5.1 poses the features of traditional analysis and contrasts them with changing characteristics. For all the centers and task forces, analysts still mostly work alone or in small groups. Their use of formal analytic methods, let alone computer-aided search engines or data-mining, is limited. Their basis for analysis is their own experience, and their tendency is to look for

Figure 5.1
Intelligence Analysis and Information Types

- Intelligence analysis is a cross-cutting discipline
  - Multiple types of information
- A central and unique intelligence community role is knowing "their secrets"
- Broader sources of information (in variable amounts) must be integrated for context

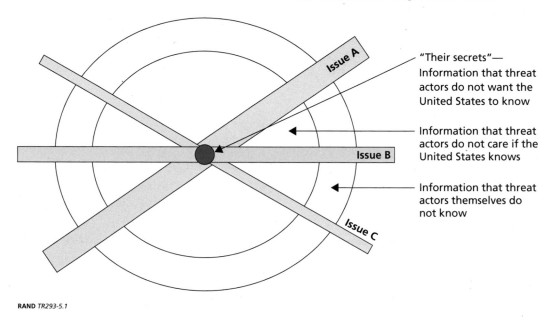

"Their secrets"—
Information that threat
actors do not want the
United States to know

Information that threat
actors do not care if the
United States knows

Information that threat
actors themselves do
not know

RAND *TR293-5.1*

information that will validate their expectations or previous conclusions. This tradecraft of the 1970s and 1980s made considerable sense when there was one over-arching target—the Soviet Union—considerable continuity, and too little information, most of which came from intelligence's own sources and thus could be regarded as reliable (the vagaries of spies notwithstanding). In the 1990s and later, however, the world of intelligence analysis was turned upside down. There were many targets, not one.[1]

An enormous amount of information exists, not all of which is readily at hand but more of which is in the "outside world" than in the world of intelligence. It is for the most part not secret, even if it is not readily available. Much of it resides not on paper but in the heads of analysts, Wall Street traders, Muslim clerics, and so on. That information varies wildly in reliability. Intelligence analysts are now their own collectors, as they search databases, reach out to expertise beyond the government, and validate what they find. Today, there is more "wheat" available but there is also vastly more "chaff." And still, the secrets of poential threats remain extremely hard to obtain.

With so much information, the use of formal methods and machine aids will increase in the future. Those methods will be used to test hypotheses and search for the out-of-the-ordinary. They will also track and remember what analysts discard and what they are watching.

---

[1] See, for instance, Treverton (2001), p. 102ff.

Table 5.1
Changing Tradecraft Characteristics

1970s and 1980s	1990s and 2000s	Future
Space for longer-term thinking	Bias toward current intelligence	Both immediate question-answering and deeper analysis
Dominated by secret sources	Broader range of sources, but secrets still primary	Draws on a wide variety of sources, open and secret
Hierarchical	Still hierarchical, though "problem-oriented-centers" added	Flat, problem-centric networks
Analysts are separated from collectors	Analysts are also their own collectors	Analysts are their own collectors
Analysts mostly passive recipients of information	Limited searching of available data	Much more aggressive searching and reaching for data, both classified and unclassified
Many analysts are deep specialists	Many, perhaps most, analysts are generalists	Mix of generalists and deep specialists, both technical and political
Analysts mostly work alone or in small groups	Analysts mostly work alone or in small groups	Analysts work in larger virtual networks
Analysis focuses on previous patterns	Same, though growing interest in new methods and tools for shaping, remembering, and examining hypotheses	Formative pattern-recognition and data-mining capabilities; searches for out-of-the-ordinary
Limited use of formal method and technology—analysts mostly operate on the basis of their own experience and biases	Limited use of formal method and technology	Wide use of method and technology—from aggregating expert views to searching, data-mining, pattern recognition
Key analytic choices with analysts	Key analytic choices with analysts	Key analytic choices with analysts
Time pressure is persistent but low intensity (mostly)	Time pressure drives toward premature closure	Technology allows memory even of hypotheses, data rejected by analysts
Institutional memory mostly in analysts' heads	Institutional memory mostly in analysts heads	Technology helps to notice what analysts are watching and asking

Analysts will operate not alone or in small groups but sometimes in larger virtual networks, as they are beginning to do today.[2] The 1970s and 1980s kept information in tight compartments, but the future will put a premium on sharing it.

Although the analytic community can benefit from new tools for analysis in a world awash in data, it is likely that the main obstacles to improved analysis in the future will continue to be not the lack of tools but, rather, the lack of infrastructure and the presence of many impeding (but few enabling) policies. The change in the nature of the threat needs to drive

---

[2] For a discussion of the need for new forms of analysis, especially against the terrorist target, see Fishbein and Treverton (2004).

what amounts to a new paradigm in analysis.[3] That means not only more use of tools and new ways of grouping and linking analysts. It also means experiments that cut against current personnel practices. For instance, cognitive psychologists tell us that creativity occurs when people are half relaxed, not running fast. Yet intelligence analysts today are running full speed all the time. So the challenge would be to create small groups for critical issues, such as how is al Qaeda morphing, and give them time and license to brainstorm, consult outsiders, and take walks, only producing when they had something to say. In this and other respects, the change in paradigm means learning the best practices from high-performance organizations.

Issues for the Future

This concluding chapter reiterates the most important issues that emerged in the report. It moves, generally, from more specific issues to broader ones. In a number of places, initiatives begun after the December 2004 legislation and creation of the DNI have moved in the directions we suggested, and we note that. For specific issues, we make relatively specific suggestions. Many of the broader ones, however, are too sweeping or fundamental to be susceptible to specific actions. Those we strive to pose as sharply as possible as subjects for further thought and investigation.

Issue 1: Workforce Data

Existing data and data calls, especially the ARC, provide a starting point, yet more needs to be done. It remains difficult to take a strategic view of analysis across the Intelligence Community, still more so to know how the Community's analytic resources align with the National Intelligence Priorities Framework. To be sure, allocating analysts to accounts is tricky because in most analytic environments, analysts handle many "issues." Yet the Community needs to be able to look over the analytic cadres in terms of numbers and analytic function, "account," experience and training levels, and so on.

Suggestion

The existing ARC should become a real data tool for taking a strategic view of the analytic community. That would make some headway in knowing how the Community lines up against priorities, but it would also be useful in identifying shortfalls in capacity, identifying training needs, and thinking about organizational forms. Community-wide agreement on data formats and update frequencies must be reached and quarterly trends relayed back to the Community as an accurate and actionable profile of the Community.

Issue 2: R&D on Methods, Datasets, Analytic Techniques, and Tools

It cannot be repeated too often that R&D for the analytic community is not just, or even primarily, about technical tools. It is about training and reshaping the culture of organizations and finding better ways to connect machines to human analysts and those analysts to one

---

[3] James Bruce uses that language of "paradigm." See Bruce (2004).

another. Here, the primary concern is connecting tool-building to the needs of analysts. The two processes are separated and often not so much at cross purposes as simply operating in different worlds. As a result, analysts often have little sense for what is available and little ability to get it if they did.

Suggestions

First, a set of basic software tools common to the analytic community would facilitate joint analysis. An analyst from one agency could move to a task force or another agency and be able to "plug and play," without first having to master a new analytic workstation. Second, and more important, some central clearinghouse would be necessary to inventory what tool-building is afoot and to connect it with what analysts need and want. Numerous non-NFIP agencies, such as DARPA and the Services, have R&D programs with elements of value to improving analysis within the Intelligence Community. Unfortunately, there are few avenues useful to collaboration during the development phases and fewer "on-ramps" that developers may take to bring their products into the Intelligence Community. The In-Q-Tel example, although quite good for the CIA, is too narrow in scope to have Community-wide benefits.

On balance, the desirability of a Community-wide perspective probably argues against what might otherwise be tempting—encouraging the major analytic organizations to consider seeking their own budgets for R&D on processes, tools, and methods. It is true that without budgets, analysts will continue to be at the mercy of what they can cajole from the builders. On the other hand, even with budgets, they would still have to depend on the tool-builders, since analysts often do not know enough to know what they want. That fact puts a premium on prototypes or experiments that show them what is available while helping them sharpen their sense of what they want most and where they are prepared to make tradeoffs. For those purposes, a central clearinghouse would be very helpful. In that sense, the creation of the DDNI(A), with staff to match, should be a step in the right direction.

That said, the analytic agencies will have to be more flexible in dealing with the cohorts of young analysts they are taking in. Traditionally, Community managers have been extremely reluctant, on security grounds, to let individual analysts customize their workstations. The concerns are appropriate, but the next generation of analysts simply will not stand for being presented with a "one size fits all" set of tools.

Issue 3: Education, Training, and Tradecraft

There is a consensus among current analysts that more should be done to leverage value from the Community as a whole, in addition to strengthening individual parts. No sense of Community capability really exists. The December legislation laid the basis for a National Intelligence University (NIU), which was created.[4] It is to be small, a coordinating and ultimately evaluating body, with the actual training and education continuing to be done by

---

[4] The 2004 law does mandate that the DNI establish "an integrated framework that brings together the educational components of the Intelligence Community in order to promote a more effective and productive Intelligence Community through cross-disciplinary education and joint training." Section 1042.

agency schoolhouses. In that sense, the NIU would be akin to the University of California system, in which the chancellors of the main campuses have considerable discretion.

Considerable value would accrue from a standard curriculum component that could be embedded in every Intelligence Community component school nationwide. The DDNI(A) began an introductory course on intelligence for the Community. A career education system similar to the Joint Professional Military Education system of the U.S. Armed Forces is a demanding model but the correct one.

Suggestions

As a first step, the analytic community should consider developing a basic course on intelligence analysis, one that would be jointly or commonly taught in the various schools. Military officers en route to being spymasters now train with CIA Operations officers, so more cooperation in training is feasible. In analysis, a common course such as Intelligence 101 might range in length from several days to several weeks. It would look at different kinds of data and reliability, outline basic patterns of inference, examine the obstacles that arise based on cognition and small-group process, and look at different kinds of intelligence problems (puzzles versus mysteries, tactical versus longer term) as well as different forms of intelligence analysis. It would outline the different forms of intelligence and intelligence analysis and what they can do. Ideally, something similar would be repeated at, say, the five- and 10-year mark in analysts' careers, when the emphasis on different analytic agencies and their forms of analysis could be much deeper.

Beyond training, the analytic community badly needs a common focal point for assessing and developing tradecraft. That probably should be a virtual center, perhaps managed by outside consultants, not a bricks-and-mortar operation. Ultimately, that virtual center might become part of a National Intelligence University that was both virtual and physical. Or, the community might delegate one agency or school as the lead. The Kent School, for instance, assisted the FBI in developing its intelligence tradecraft. This "center" might conduct activities along two lines. The first would be more systematic enquiry into particular problems of tradecraft—like the Kent Center's recent work on the particular features of analyzing transnational issues, such as terrorism, in contrast to more traditional state-centric issues.[5]

The other would be to serve as the provocateur and seedbed for a variety of experiments with new methods and tools, especially but not only those that seek to combine analysts with machines in ways that capitalize on the special strengths of both. Finally, nurturing the value of exploring competing hypotheses and developing a path for their use in analysis is important but not simple. This is notionally illustrated in Figure 5.2.[6]

In these regards, too, there are the beginnings of progress, if not exactly in the form of our suggestions. Pushed by the December legislation and the subsequent report of the WMD Commission, the DNI has moved to create a Longer-Term Analysis Unit in the National Intelligence Council, and a DNI-managed Open Source Center as a seedbed for making better and more creative use of open sources in analysis.

---

[5] See Fishbein and Treverton (2004).

[6] Poindexter (2002).

Figure 5.2
Plausible Futures and Actionable Options from Competing Hypotheses

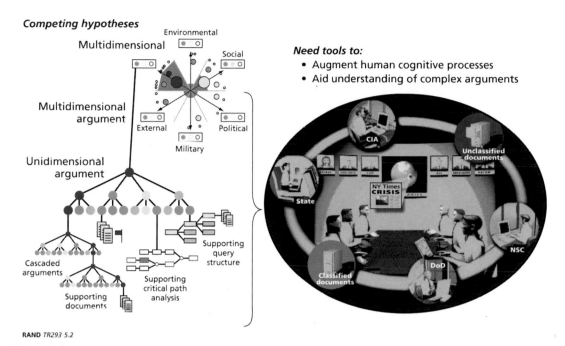

RAND TR293 5.2

The lower right-hand side of this figure pictorially indicates the cycle of analysis in a way intended to reflect the critical balance between human cognition and the digital information environment (databases, retrieval and processing methods, etc.) that is needed to support effective future analytic tradecraft. Tradecraft needs to reflect both the spirit and the substance of analyzing competing hypotheses. Multidimensional information must be synthesized from unidimensional arguments developed out of a range of different perspectives, including environmental, military, political, social, and others. The most compelling competing hypotheses (shown in as red dots in Figure 5.2) then would be carried forward from subordinate analytic processes, such as cascaded arguments, query structures, and critical path analyses, into higher-level refined alternatives to the initial interpretations of analysis. These competing hypotheses could also provide the basis for both field experiments and scenario-based games to test the logic, consistency, and credibility of emerging policy positions.

The policymaking and operational communities would take the hypotheses represented by models from the analytic community, estimate plausible futures, and create actionable options for the decisionmaker.

Issue 4: Getting and Keeping the Next Generation of Analysts

Compensation reform is an opportunity for the Community to reward people for their work, not their seniority. The challenge will be to make sure that perverse incentives, such as still more disincentive to longer-term deeper analysis, do not arise in the process. So, too, will the incorporation of the next generation of analysts be a great opportunity for the Community, for much of what these new analysts will demand will be directions the analytic community should be moving toward in any case.

Suggestions

Most of the intelligence agencies have already tried monetary and other immediate incentives to attract a new generation of top-flight talent. But the harder challenge will be to reshape work patterns to meet the demands of that new generation. Better tools and more networking are part of the answer. Another part would be experiments with new work processes, more virtual groups, "chat rooms," Web logs, and ways of putting analysts who are precisely *not* the ostensible experts to work with those who are. The initiative to create an "Intellipedia" is a promising one. Many of these experiments will run into objections grounded in security. Accommodating to technology that younger people use routinely will be important to expanding their abilities to think and interact.

New ways of mentoring need also to be found. The best Wall Street firms, for instance, take advantage of "gray-green" age distributions by putting fresh and fearless younger people together with the most seasoned older heads.[7] And the Community will have to find ways to respond to the restlessness of many newcomers, who will want to develop real expertise but will be turned off by being relegated to a "few square miles of Iraq" as their principal assignment.

Issue 5: Data-Sharing, Compartmentation, and Secrecy

There is a continuing concern that analysts are not being given the data they need to do their jobs. They are being given processed intelligence, not source material, and they see data owners and processors becoming confined to single sources, unable to take a broader view and to know what other sources might offer. NSA, however, is now taking distinct steps to reverse this trend by beginning a number of initiatives in sharing data. Policy guidance is in place that supports data-sharing, and statements by Community leaders all publicly support these principles, but sharing does not happen. Leadership commitments regarding data ownership and sharing seem shallow—an issue that does warrant further evaluation.

One particular of this challenge was at the core of the WMD Commission findings: Analysts often know too little of the sources on which they depend, especially the human sources, to judge their reliability. This challenge, like most others, invokes a tradeoff: Giving analysts more information to judge sources carries some increased risk of disclosing those sources. On balance, though, the CIA, in particular, has decided that the risk can be managed and has taken a number of steps to better validate sources in analytic products, including having the relevant chiefs of the clandestine service present at meetings when national intelligence estimates are discussed and approved. New tagging technology also permits information to move through the system carrying indications of its provenance.

More generally, though, the culture of compartmentation and secrecy frustrates many innovations, an important issue for the Intelligence Community as a whole. The current security and technology approaches are precisely what are not needed for effective analysis. For example, for many analytic problems in this new world, the analysts who have no evident need-

---

[7] This observation was made by Roger Kubarych, a former director of research for the New York Federal Reserve.

to-know may be the most valuable in making new connections or seeing new patterns. Some of the most interesting experiments we observed, in "multi-INT" for instance, do not draw on a widely distributed enterprise but depend, rather, on place. That is, so long as they are small and experimental, they can get license to operate "within the security fence," sharing information in ways that the originating agencies probably would not have permitted on a larger scale. In one example, an analyst literally faces a handful of computer screens, and "fuses" information by rolling his chair from one to the next (what might be called "wheeled fusion").

It is interesting to note that although compartmentalization and secrecy seem to us enormous issues, surprisingly, they hardly came up in our interviews, except in the Department of Homeland Security, which has a major problem handling the combination of foreign and domestic intelligence data, especially in how to treat the names of U.S. persons—that is, citizens and resident aliens. It may be that most analysts have found what they regard as acceptable work-arounds for the most debilitating secrecy procedures or that they simply no longer notice them.

Suggestions

Sooner or later, the elements of future tradecraft will run up against the traditional approach to protecting sources and methods. Future analysis seeks to share information, not cosset it. It presumes that fresh insights will come precisely from those who do *not* have a need-to-know. For the ostensible experts, the future is overdetermined; they can cite a myriad of reasons why the future will turn out like the past. But a newcomer, one without obvious need-to-know, might see new patterns or hints of new futures. If FBI agents had paused over those Middle Eastern flight school students in the United States during the summer of 2001 who were uninterested in landing and taking off, that behavior might have seemed simply strange, not fitting with any threat profiles on their minds. Outside terrorist experts, though, might have paused longer to reflect on what that behavior might mean.

Squaring the future of analysis with the old imperatives of security will be both critical and no mean feat. In the short run, there is no alternative to a variety of experiments with new ways of sharing. Most of those, like the "wheeled fusion" described above, will have to be small, "inside the fence," and so not very threatening.

In the longer term, the problem of security ought to be less pressing, at least in principle. During the Cold War, intelligence depended heavily on a small number of collectors, so any single-point exposure was deeply damaging. Arguably, that is less so now with much more varied targets and much more information. Even if that is true, however, it still means that intelligence will have to recognize, as Silicon Valley has, that innovations that confer advantage are often fleeting. In many cases (but not all), if advantage is to be maintained, it will require a short cycle in producing innovations.

Issue 6: All-Source Analysis

The fundamental definitions of analysis have changed in ways that will require a more agile and adaptable future Intelligence Community. Take the notion of all-source analysis as an example. As we said at the outset, distinguishing it from single-source analysis may no longer make sense. Integrating information from multiple sources has always been a challenge both institutionally and substantively. Institutionally, different organizations do the work, and coordinating or integrating their efforts is frequently not easy.

Technically, since the disciplines associated with processing and analyzing particular kinds of raw intelligence data can be so different, analysts who specialize in one particular form of intelligence may not fully appreciate what others can contribute and, as a result, may not even know how to ask the right questions of their counterparts in other disciplines. Moreover, there will be times when the analyst must go outside the Community to subject matter experts in academia or industry. When and how to do this is often unclear and there are few (perhaps no) tools to help the analyst in this area.

All that said, successful integration of intelligence from different sources does occur, and when it does, it demonstrates the potential power of using different kinds of collection systems in complementary ways to produce a more complete, integrated picture. For example, in an illustration from our interviews at NASIC, information from one particular kind of intelligence collection system had provided some interesting but incomplete information about a particular matter of interest. Yet that information was sufficient to allow other different kinds of collection systems to be tasked to provide the missing pieces of the puzzle. Thus, not only was the integrated information more valuable than the sum of the parts, but also the information from one collection system was used to help target another.

This successful "horizontal integration" seems to rest on three requirements:

- A specific problem to solve;
- Different analytical groups as part of the same organization—preferably housed in the same facility—or at least analysts who know each other and have a basic understanding of what the others do; and
- Analysts able to at least influence collection system tasking; otherwise, the point of the integration is lost.

Determining how to best integrate all available information on a particular subject into a coherent analysis that answers policymakers' questions is the most important problem in intelligence analysis. However, what is commonly called all-source analysis appears to do something different in practice. First, almost by nature it inadvertently relegates the organizations that are not considered all-source to a supporting role in the analytical hierarchy. Yet many types of sources might not be required to make an assessment of a situation. This is particularly true in intelligence analysis supporting military operations. One source, SIGINT for instance, might be sufficient to fill a pressing need or at least be the dominant input. Filtering the information through all-source analysts would add little and might actually distort the product. Analysts who specialize in one particular kind of intelligence will frequently develop enough insight to understand the context and implications of their specialized information for policymakers, decisionmakers, and military leaders.

Second, all-source agencies at the top of the hierarchy can do very little to solve the horizontal integration problems mentioned above. Delegating all-source integration only to those at the top of the chain misses opportunities for improved collection targeting and more effective analysis throughout the intelligence cycle depicted in Figure 2.1. All-source is probably inappropriate even as a metaphor, because it ties everything to collection systems. What is more appropriate is informed insight, doing whatever is necessary to solve specific problems, which might or might not involve multiple kinds of intelligence. "All-methodologies," "all-disciplines," and "all-perspectives" analyses are likely to be more powerful than all-source analysis. Most policymakers simply want to know whether the information being conveyed is based on everything known about the issue.

Suggestion

Here, the suggestion is simple, but the implication is large: End the distinction between single- and all-source intelligence. Distinguishing intelligence products by the number of kinds of sources they use is wrong-headed at best, silly at worst. And be mindful of the emerging distinctions that are taking place within analysis between short-term, horizontal integration, and longer-term analysis. Although tying all intelligence data to a precise, common space-time framework (co-registration) sounds logical, it has contextual weaknesses that must be dealt with before analysis can be done. First, the aggregation of data has its own complexity and can exhibit a "weak link in the chain" characteristic to the extent that some or all of the data are fragmentary, deceptive, or even missing. (Concern over "missing something" has been exacerbated by the events of September 11th.) Similarly, although the creation of a data stack of intelligence information is potentially very important in explaining an issue or event, the quality of analysis will depend greatly on the extent to which those performing the analysis are familiar with the strengths and weaknesses of all of the datasets in the stack and generally familiar with how to do good analysis. RAND's "Day After" method and other similar exercises might be used to explore, in concrete cases, the challenges of integrating, then analyzing, information from very different sources.[8]

Issue 7: New Kinds of Intelligence, Especially "Domestic"

As Figures 2.1 and 3.1 illustrate, different forms of analysis are located in different places in the analysis cycle. National intelligence remains the endeavor by which politically, diplomatically, and militarily relevant information about potentially threatening nations, groups, and individuals—information that they do not want the United States to know—is determined, gathered, and transformed into new insights and knowledge. However, as notionally illustrated in Table 3.1, national intelligence involves more. It must also include important things that the United States needs to know about nations, groups, or individuals that they do not care that we know, and it may involve things that they really do not know themselves. This transformation of gathered information into high-level, new knowledge and global awareness is intelligence analysis. That analysis continues to be a tradecraft drawing on insights, secrets, and on-the-job experience rather than taught, academic disciplines. It must address both mysteries as well as puzzles, as defined above.

---

[8] "Day After" gaming gives participants enough background to play through a future crisis or other set of choices, then takes them from the "day after" back to today, encouraging them to ask: Having seen one future, what should we be doing today? In this case, how should intelligence be done to better shape the basis for decisions tomorrow?

Yet, all the familiar components of the intelligence cycle illustrated in Figure 2.1—national, national foreign, S&T, and military—are being reconfigured and combined with increasingly important forms of intelligence, such as "targeting analysis" or "law enforcement intelligence," which are being highlighted by today's challenges.[9] The line that formerly separated intelligence and law enforcement is being blurred, driven by transnational threats such as terrorism. In the process, both the FBI and DHS will become more central members of the Intelligence Community—both a challenge and an opportunity for the future. Law enforcement officers, even in their traditional roles, are collectors who could be much more useful to the broader community. Making them so will require not just technology and analytic methodologies but also infrastructure to guide them and retrieve what they collect. And that, in turn, will require changes in national policy.

A beginning to this reconfiguration was the creation of the FBI National Security Branch in 2005, which the WMD Commission had recommended, bringing together the FBI's counterterrorism, counterintelligence, and intelligence functions, with a strong oversight role by the DNI. In effect, the nation decided, for the time being at least, not to create a separate domestic intelligence service, a version of Britain's MI-5.[10] Instead, it opted to let the FBI continue with its effort at reshaping, from law enforcement to prevention and intelligence.

Suggestion

Here, the suggestion is a crucial issue to consider. Beyond the FBI National Security Branch, does the nation really need to distinguish between domestic and foreign intelligence functionally and organizationally, not to mention legally? As there is more legal room to mix the two, what does that mean for organizations? What new pressures will build and how will perceptions change regarding domestic and foreign issues *following the next September 11th?*

A more thorough assessment of this foreign-domestic issue is warranted before organizational change is accomplished. That assessment should include the historical rationales for creating the CIA in the first place, and then, later, for establishing a line between intelligence and law enforcement, including within the FBI itself. Since September 11th, that line—or "wall" in FBI parlance—has been all but eradicated. Surely, intelligence and law enforcement are being pushed together in the war on terrorism, and so the issue is how and how fast to push that embrace or ratify it in organization. Another argument against the MI-5 solution was that having torn down the wall between domestic intelligence and law enforcement, it made little sense to rebuild the wall by creating a separate domestic intelligence service.

Issue 8: Analysis versus Collection-Driven Intelligence

Is the future really "analysis-centric" instead of "collection-centric," as discussed in Chapter 2? Most of our interviewees thought so. Again, the answer to this question may differ across agen-

---

[9] The lack of a real intelligence function at the FBI and the ragged cooperation between intelligence and law enforcement in general, and the FBI and CIA in particular, are central themes of all the September 11th post mortems. See, for instance, the *9/11 Commission Report* (2004).

[10] For a discussion of arrangements in four other democracies, see Chalk and Rosenau (2004).

cies. But the answer has implications throughout the intelligence cycle, from collection strategies to alternative interpretations. And while the catchphrase "analysis-centric" is in vogue, so much open-source analysis is being done by think tanks, universities, nongovernmental organizations (NGOs), and private companies that *perhaps the Intelligence Community's future is precisely coupled to collecting those secrets for issues where secrets matter most.* "Secrets" here would refer more to secrets of substance than to secrets of sources and methods. To do such collection, the Intelligence Community must function as the only enterprise in a position to create a unified picture of events and issues based on both open-source information and secrets.

Certainly, if the Community organized more around specific problems—in centers such as the National Counterterrorism Center—collection necessarily would be driven more by analysis needs than by collection capabilities. However, the best balance between problem-driven collection and search collections cannot be determined a priori. A floor must be established for search collections at some predetermined percentage utilization of resources to hedge against surprise, while maintaining enough collection horsepower to target specific tasks that will answer analysts question on current issues. Collecting information that is never used further is clearly not wise; in an era when the Community is awash in too much unused data, collecting more makes even less sense. In the end, the Community's inability to analyze the data it collects may be a greater risk than any shortfalls in its performance.

Suggestion

As more analysis is organized around problems, ideally in teams that include collectors, both the opportunity and the knowledge to drive collection will grow. Indeed, it may be necessary for collection managers to set some floor of the sort described above. Recent collaborations between the Community's central managers for analysis and for collection also seem to have been fruitful, and more of those make sense—particularly for countries, issues, or problems of enduring interest that may not be currently "hot" and so lack champions in bidding for collection. However, it does seem apparent that production and distribution issues involve equities quite different from analysis and are likely best treated separately.

Issue 9: Reporting versus Analysis

If the fundamental issue for the future is defining analysis, a related issue is the balance between longer-term analysis and short-term reporting. This report has underscored the pervasive concern that responding to the "crisis of the day" has had a negative effect on the Community's ability to maintain a core with relevant subject matter expertise in the analytic workforce. This pressure is exacerbated by the absence of a focused threat comparable to that of the Soviet Union and its weapons acquisition system, which galvanized the Community and nurtured the value of deep expertise, built up over years of dedicated operational and scientific analysis effort.

This tension is felt in the analytic community with pressures for PDB-like reporting to understand, forecast, and counter terrorism, for example, and in the military Intelligence Community, where the drive toward the force-protection needs of current conflicts often overshadows scientific and technical challenges of determining capabilities in regions not currently involved in combat. The press of the immediate is hardly new, and many analysts, across agencies, lament the relative lack of deeper analysis, but most also see little interest in such work from consumers, perhaps especially those now in office. Again, the answer may differ across agencies, from CIA to INR, from NCTC to CTC, but again, too, the question of opportunity cost arises. Although reporters can be experts, and sometime even report in their area of expertise, the fact is that reporting well is rewarded and deep expertise is generally not.

A related issue is whether the analytic community is in the information business or the secrets business. Is the Community the integrator of information for policy or the purveyor of secrets plus context? What is the right balance, within agencies and across them, between longer-term analysis and short-term needs? Again, this issue raises many more far-ranging issues than opportunity cost questions alone and demands a more careful and precise definition of "secrets."

Suggestions

There is no point of authority or assessment process now in place to oversee the balance between intelligence and tradecraft-building efforts for long-term needs and responsiveness to short-term priority areas. The creation of the Longer Term Analysis Unit at the NIC is a step in the right direction, but its success will turn on how much it can promote longer-term analysis around the Community, not just become a long-term analytic ghetto. Beyond that, there should be a focal point at the DDNI level that can evaluate the consequence of creating a center for this or that on broader intelligence capabilities and longer-term capacities needed for the future. By our lights, this problem should be solvable. Creating a dedicated cadre of long-term analysis to focus on key problems, and allowing for the timely review and dissemination of their product while keeping them loosely linked to the producers of current intelligence, seems eminently doable. These two aspects of analysis will also benefit from key individuals at multiple seniority levels being rotated between them, underscoring the importance of both.

Issue 10: Organizing the Intelligence Community

The narrower issue of how analysts are deployed merges into the larger one about how the Community is organized. In the first stages of processing the information that is collected, the stovepiped organization for the various INTs is acceptable and perhaps essential. This is the initial analysis level, including both technical data and HUMINT information collection and data-processing. Even at that stage, though, the stovepipes may miss opportunities to shape collection around problems and include broader sources. As analysis begins to change processed data into information, however, and intelligence products are being developed from multiple sources, an INT-specific organization is not appropriate. Be it problem-centric, customer-centric, or something else, a more horizontally distributed Intelligence Community must be functionally forged as an intrinsic part of the analytic tradecraft process. It must be structured to make the best use of tools that are constantly evolving and changing and tailored to the development of tradecraft, not production.

Suggestions

A distributed organization is essential here, one that can reach to subject matter expertise worldwide, both inside and outside the Intelligence Community, as needed to support analysis quality and debate alternative interpretations. Additionally and differentially, the Community may be able to function as a virtual enterprise with cells centered on problems or themes (or regions). Figure 5.3 illustrates that a distributed, virtual community for analysis may be able to address the multiple community-wide issues outlined above. Shown in the upper-left portion of the figure is a distributed intelligence enterprise with the DDNI(A) hub at the central core. This central hub is tied to other hubs involving the DoD, the DoE, the Intelligence Community, and related intelligence agencies with high-data-rate links, each of which is linked to its component center of excellence as well as to individual subject matter experts inside and outside academia and industry. Relevant components not inside the traditional Intelligence Community are shaded in gray in the figure.

Figure 5.3
A Virtual DDNI R&D Enterprise

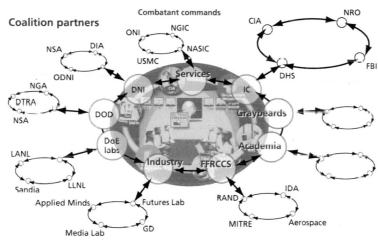

- DDNI creates a bundle of three to five imitative centric "tradecraft" cells
- Each cell involves
  - Multiple agency participants
  - Tailored age/experience ranges
  - Themes tailored to tradecraft needs
- Each cell bundle accesses a hub, digitally linked to pertinent expertise worldwide *and* performing cooperating partners in industry and academia

Outside experts

- Cell personnel *both execute and manage* tradecraft-related activities in specialized theme areas
- Cell initiative themes may be:
  - Community-wide training and education related
  - Hypothesis and alternative analysis testing
  - Problems nurturing global analysis

NOTES: ODNI = Office, Director of National Intelligence; DTRA = Defense Threat Reduction Agency; LANL = Los Alamos National Laboratory; LLNL = Lawrence Livermore National Laboratory; GD = General Dynamics; IDA = Institute for Defense Analyses; ONI = Office of Naval Intelligence; NGIC = National Ground Intelligence Center; NASIC = National Air and Space Intelligence Center.
RAND TR293-5.3

The DDNI(A) is shown as a central hub, which would enable the office to act as a coordination point for analysis R&D throughout the Community. In fact, any of the hubs shown could assume the role of a master hub, depending on the nature of the problem or issue to be addressed. As noted in the figure, the DDNI might establish a set (or bundle) of problem-centric or initiative-centric subnetworks within this architecture, each with its own master and support hubs chosen to fit the problem of interest. Analysis of crosscutting problems or theme-areas form the National Intelligence Priorities Framework (NIPF). Issues such as terrorism, WMD proliferation, or regional challenges such as China may be more effectively examined using this topology than with the current hierarchical and stovepiped organization. This "slicing" of the hierarchical organization into problem-centric virtual sheets, unified by a problem or interest area, is illustrated in the lower-right corner of Figure 5.3.

Organizational structures are hierarchies or networks. Hierarchies are important, but the "work gets done" and trust is built in networks. One must create shared workspaces that facilitate the operation of "trust-networks" within organizational hierarchies to bring diversity to problem solutions where uncertainty exists.

A distributed enterprise can be immensely valuable in addressing a number of challenges facing the analytic community of the future. This would be all the more so if that enterprise were, in turn, enabled by a distributed, high-speed network that linked to subject matter experts worldwide, to both deep and broad analysts within and outside the Intelligence Community, to cutting-edge technology developers in industry, and to visionaries in academia and elsewhere. Example partnerships might include academia (e.g., Media Lab), industry (e.g., Futures Lab[11]), and government (e.g., NRO/AS&T/FL) for specific tradecraft RDT&E roles.

One attractive feature of this enterprise would be its ability to function as a virtual organization, pulling teams together into cells, formed around problems, experiments, and educational test beds, and then disbanding the virtual workforce when the activity is complete. The infrastructure to do this is basically in place now and minimal investment is likely required. The list below provides a few examples of the problem-centric cells that would best be supported by this type of enterprise:

- Global challenges in areas not driven by current priorities;
- Alternative analyses and hypotheses testing;
- Experiments in law enforcement collection analysis;
- Topical experiments on all-source versus limited source analysis; and
- Lessons learned from key intelligence successes and lapses.

A better understanding of the economic models underpinning a virtual community (such as how the cost of multiple analysts from multiple agencies, dedicated to a particular community-wide issue, are reimbursed to the home agencies) must be developed before this organizational option can be seriously considered.

---

[11] Futures Lab, a virtual analytic environment serving the needs of the Intelligence Community in general and the analytic community, particularly, is part of the National Reconnaissance Office's Advanced Systems and Technology Directorate. It is housed at NRO headquarters and has worldwide secure research.

A Final Word

In the technology arena, development of a nationwide (ultimately global), high-speed digital infrastructure able to continuously support gigabit data rates is more important than analytic software tools.[12] Those tools, to be sure, are important, but they will continue to be developed in the commercial community as processing power and technologies continue to evolve. Government nurturing is not likely to be needed. Infrastructure needs and opportunities will require larger investments. Both leadership and funding are essential and continuity in both is difficult within the government. More important than R&D and technology development per se is a better understanding of the role of the Intelligence Community and consistently applied metrics for rewarding those who excel. Education and training standards, processes, and costs may then be identified, budgeted, and executed in a balanced manner that can nudge the workforce as a whole in a positive direction.

None of these issues, however, is as important to the future of analytic tradecraft as is the quality of Community leadership. In simple terms, the national and community leadership devalues intelligence analysis today, and the analytic community is aware of this. Policies and proclamations abound that endorse the importance of intelligence analysis, data-sharing, fusion priorities, and the like, but the will and intent to enforce them carries a political, cultural, and social price that is simply viewed as too high for the likely results. This must be changed if tradecraft is to serve the nation better in the future.

The exhortation is general, but the suggestion is specific: Build the DDNI(A) as the hub for a Community-wide perspective on goals, training, and tradecraft. That hub would be virtual, distributed, and federated—virtual because no new organizations are necessary or feasible; distributed because technology offers leverage while reducing costs; and federated because the analytic agencies have to own the program. The hub would nurture a number of cells, for instance in Community-wide training and education; in testing hypotheses and alternative analysis; and in nurturing global analysis. Each cell would combine analysts from different agencies, of different specialties and career stages, all oriented toward tradecraft needs. They would, in short, take advantage of the richness of the Intelligence Community's analytic abilities.

---

[12] Albert (2003).

CHAPTER SIX

Recommended Actions

Establish DDNI(A) as a focal point to evaluate opportunity costs and assess "right balance" in analysis
— Collection-driven versus analysis-driven
— Current reporting versus longer-term analysis
— In-house versus outsourced

Foster better integration of methods and tools for analysis
— Establish focal point to connect R&D and tool-building community (government and industry) to Intelligence Community analysts
— Develop minimum common tool set for community-wide use

Institute community-wide tradecraft training and education components
— Develop tradecraft curricula for community-wide use
— Institutionalize lesson-learning as process of performance improvement, not assessing blame

Get and keep the next generation of analysts
— Build partnerships with academia (e.g., Media Lab), industry (e.g., Futures Lab), and government (e.g., NRO/AS&T/FL) and link new hires
— Track promotion, retention, and erosion rates for new hires over decade
— Align training, incentives, processes, and metrics with performance

Innovate in analytic methods and data-sharing
— Promote a variety of experiments and field tests, mostly "inside the security fence," as demonstrations and validations
— Recognize that the nature of secrecy is changing

Evaluate the boundaries of all-source versus single-INT analysis
— End the distinction at mid and high levels of analysis; analysis is not distinguished by the number of sources
— Develop portfolio of "Day After" games, and other simulations, to nurture transitions

Rethink new kinds of intelligence, especially law enforcement
— Focus on usefulness, necessity of "domestic/foreign" divide
— Use gaming to explore gray areas

Bibliography

Alberts, David S., *Information Age Transformation: Getting to a 21st Century Military*, Washington, D.C.: CCRP Publication Series, 2003.

Bruce, James, *Dynamic Adaptation: A Twenty-First Century Intelligence Paradigm*, Central Intelligence Agency, 2004.

Business Executives for National Security, Pay for Performance at the CIA: Restoring Equity, Transparency and Accountability, January 2004. As of April 25, 2007:
http://www.bens.org/images/CIA_Reform%20Report.pdf

Carmel, David, "Experiments in TREC—the World Championships in Search Engines," 2004. As of April 25, 2007:
http://www.research.ibm.com/haifa/Workshops/summerseminar2004/present/trec_summary.pdf

Chalk, Peter, and William Rosenau, *Confronting "The Enemy Within": Security Intelligence, the Police, and Counterterrorism in Four Democracies*, Santa Monica, Calif.: RAND Corporation, MG-100-RC, 2004. As of May 24, 2007:
http://www.rand.org/pubs/monographs/MG100/index.html

Dervarics, Charles, "Smart Maps: Geospatial-Intelligence Visualization Tools Speed Planning, Analysis and Collaboration," *C4ISR Journal of Net Centric Warfare*, March 1, 2005. As of May 24, 2007:
http://www.isrjournal.com/story.php?F=617293

Dewar, James, *Assumption-Based Planning: A Tool for Reducing Avoidable Surprises*, Cambridge, UK: Cambridge University Press, 2002.

Final Report of the Commission on the Intelligence Capabilities of the United States Regarding Weapons of Mass Destruction [WMD Commission Report], Washington, D.C., 2005. As of September 2005:
http://www.fas.org/irp/offdocs/wmdcomm.html

Fishbein, Warren, and Gregory F. Treverton, *Making Sense of Transnational Threats*, Central Intelligence Agency, Kent Center for Analytic Tradecraft, Occasional Papers, Vol. 3, No. 1, October 2004. As of March 2, 2005:
www.cia.gov/cia/publications/Kent_Papers/pdf/OPV3No1.pdf

Freedman, Lawrence, *U.S. Intelligence and the Soviet Strategic Threat*, 2d ed., Princeton, N.J.: Princeton University Press, 1986.

"Global Awareness," *Essays on Air and Space Power*, Vol. 1, No. 8, Maxwell Air Force Base, Ala.: Air University Press, 1997. As of May 24, 2007:
http://www.cadre.maxwell.af.mil/ar/MENTOR/vol1/sec08.pdf

Intelligence Reform and Terrorism Prevention Act of 2004. As of January 4, 2005:
www.fas.org/irp/congress/2004_rpt/h108-796.html

Janis, Irving L., *Victims of Groupthink: Psychological Study of Foreign-Policy Decisions and Fiascoes*, Boston, Mass.: Houghton Mifflin, 1972.

Johnston, Rob, *The Culture of Analytic Tradecraft: An Ethnography of the Intelligence Community,* Washington, D.C.: Center for the Study of Intelligence, Central Intelligence Agency, 2005.

Kahn, Herman, et al., *The Next 200 Years—A Scenario for America and the World,* New York: Leon Martel, 1976.

Kent, Sherman, "Words of Estimative Probability," in Donald P. Steury, ed., *Sherman Kent and the Board of National Estimates: Collected Essays,* Washington, D.C.: Center for the Study of Intelligence, 1994.

Lempert, Robert J., Steven W. Popper, and Steven C. Bankes, *Shaping the Next One Hundred Years: New Methods for Quantitative, Long-Term Policy Analysis,* MR-1626-CR, Santa Monica, Calif.: RAND Corporation, 2003. As of May 24, 2007:
http://www.rand.org/pubs/monograph_reports/MR1626/index.html

National Commission on Terrorist Attacks Upon the United States, *The 9/11 Commission Report,* Washington, D.C., 2004. As of April 25, 2007:
http://www.9-11commission.gov/

Nye, Joseph S., Jr. "Peering into the Future," *Foreign Affairs,* Vol. 77, No. 4, July/August 1994, pp. 82–93.

Percivall, George, and John Moeller, "Integrating Intelligence," *GeoIntelligence,* Vol. 1, November 2004. As of May 24, 2007:
http://www.geointelmag.com/geointelligence/article/articleDetail.jsp?id=134279

Poindexter, John, "Total Information Awareness, DARPA's Initiative on Countering Terrorism," presentation at DARPA TECH 2002, Anaheim, California.

Prados, John, *The Soviet Estimate: U.S. Intelligence Analysis and Russian Military Strength,* New York: Dial Press, 1982.

"Report of the Independent Commission on the National Imagery and Mapping Agency," Washington, D.C., January 2001. As of April 12, 2005:
http://www.fas.org/irp/agency/nima/commission/article05.htm#3

Senate Select Committee on Intelligence, "Report on the U.S. Intelligence Community's Prewar Intelligence Assessments on Iraq," July 2004. As of May 24, 2007:
http://www.fas.org/irp/congress/2004_rpt/index.html

"Tasking, Processing, Exploitation & Dissemination (TPED) TPED Analysis Process (TAP)," GlobalSecurity.org. As of May 24, 2007:
http://www.globalsecurity.org/intell/systems/tped.htm

Treverton, Gregory F., *Reshaping National Intelligence for an Age of Information,* Cambridge: Cambridge University Press, 2001.

Tufte, Edward R., *The Visual Display of Quantitative Information,* Cheshire, Conn.: Graphics Press, 2001.

Van der Heijden, Kees, *The Sixth Sense: Accelerating Organizational Learning with Scenarios,* New York: John Wiley, 2002.Tufte, Edward R., *The Visual Display of Quantitative Information,* Cheshire, Conn.: Graphics Press, 2001.

Van der Heijden, Kees, The Sixth Sense: Accelerating Organizational Learning with Scenarios, New York, John Wiley, 2002.